THE DISCOVERY OF EVOLUTION

The Discovery of Evolution explains what the theory of evolution is all about by providing a historical narrative of discovery. The narrative shows how some of the major puzzles that confront anyone who studies living things were solved from an evolutionary perspective. Beginning with the work of naturalists in the seventeenth century, the scientific discoveries that led up to, and then flowed from Darwin and Wallace's theory of evolution by natural selection are explained. This new edition of the *Discovery of Evolution* is fully updated and contains a new chapter on the evolutionary studies of the late twentieth century. By approaching the topic of evolution in this way, it is made accessible to the non-specialist and no previous study of biology is required.

DAVID YOUNG is Associate Professor in the Department of Zoology at the University of Melbourne, Australia.

THE DISCOVERY OF
EVOLUTION

2ND EDITION

David Young

CAMBRIDGE
UNIVERSITY PRESS

IN ASSOCIATION WITH
NATURAL HISTORY MUSEUM, LONDON

CAMBRIDGE UNIVERSITY PRESS

Cambridge, New York, Melbourne, Madrid, Cape Town, Singapore, São Paulo

Cambridge University Press
The Edinburgh Building, Cambridge CB2 8RU, UK

Published in the United States of America by Cambridge University Press, New York

www.cambridge.org
Information on in this title: www.cambridge.org/9780521868037

First published 2007

Printed in the United Kingdom at the University Press, Cambridge

A catalogue record for this publication is available from the British Library

Library of Congress Cataloguing in Publication data

Young, David, 1942 Feb. 25-
The Discovery of evolution / David Young. – 2nd ed.
p. cm.
Includes bibliographical references.
ISBN-13: 978-0-521-86803-7 (hardback)
ISBN-13: 978-0-521-68746-1 (pbk.)
1. Evolution (Biology)—History.   I.   Title

QH361.Y695 2007
576.809–dc22

2007002768

ISBN-13   978-0-521-86803-7   hardback
ISBN-13   978-0-521-68746-1   paperback

CONTENTS

*Colour plate section appears between pages 56 and 57, 88 and 89,
120 and 121 and 152 and 153*

'Reading the history of evolutionary biology is one of the best ways to understand what evolution is and what it means for understanding our place in the world. *The Discovery of Evolution*, beautifully illustrated with contemporary pictures and documents, shows how the theory of evolution provided a single, elegant and convincing answer to a myriad of biological questions. Taking the reader from Aristotle to Darwin to Hamilton and beyond, this book is an accessible introduction to the development of evolutionary biology, and should be read by anyone interested in how this revolutionary idea changed the way we think about the world around us.'

Lindell Bromham, School of Botany and Zoology, Australian National University.

INTRODUCTION

I invite you to accompany me on a journey of adventure and discovery. What I offer you here is a journey for the mind and adventure in the realm of ideas. In the following chapters, we shall go back in time and retrace the steps of the scientists who developed the theory of biological evolution. We shall follow the thoughts of those who first made discoveries relevant to the theory and those who grappled with evolutionary ideas when they were new. This will enable us to see what led scientists to accept that the Earth was far older than previously thought, to prefer evolution to special creation and to grasp the nature and importance of natural selection.

No one need hesitate to set out on this journey on the grounds that they know little about biology in general or evolution in particular, for our journey begins in the seventeenth century, when even the most accomplished naturalists knew next to nothing about biology as we understand it today. By the seventeenth century, the rebirth of learning in Europe had kindled a renewed curiosity about the natural world of animals, plants, minerals and rocks. The naturalists of the time carried out their work with great enthusiasm but with only the simplest of scientific backgrounds to guide them.

As we shall see, their steadily increasing knowledge brought naturalists of the seventeenth and eighteenth centuries up against some of the fundamental problems of biology. One thing that they encountered was the great diversity of animals and plants found across the world. Another was the occurrence of fossils embedded in the rocks, a discovery that opened up the history of life. Comparative anatomical study began to reveal the detailed organization by which animals and plants are adapted to their particular way of life. Dealing with this expanding field of knowledge required new ideas and methods, and these developments led to the origin of biology and geology as professional fields of study.

The terms 'biology' and 'geology' came into use at the beginning of the nineteenth century, which was a vitally important period for the development of evolutionary theory. The quickening pace of research early in the century began to uncover much new evidence and, in the middle of the century, Charles Darwin and Alfred Wallace incorporated this evidence in a comprehensive theory of evolution. This theory was promptly taken up by men like Thomas Huxley and was soon accepted into biology and geology at large. By the end of the century, it was clear that Darwin and Wallace had brought about a permanent advance in our understanding of the living world.

The nineteenth century is thus the high point of our journey, and fortunately it is an accessible and fascinating period. People jotted down their thoughts in notebooks and shared their thoughts by letter in those days before the advent of telephones or email. These notebooks and letters were often carefully preserved and, for a number of famous scientists, they have been published in book form. Many of these are still readily available, along with the regular books and articles written by these people. So, to this day, I have on my shelf more books *by* Darwin, Huxley and Wallace than books *about* them.

Our journey then moves on to the twentieth century, where biology becomes steadily more diversified and specialized. Consequently, it is not possible to take as close a view as we obtained of the pivotal episodes in the nineteenth century. We shall instead take a path that picks out some of the crucial developments in the study of inheritance, of populations and of fossils, which have progressively refined the theory of evolution. As in any healthy field of science, these developments generated a stimulating mix of confusion and clarification, controversy and synthesis. But they also confirmed and strengthened evolutionary theory, establishing it as a crucial ingredient of modern biology.

The outcome of taking this path through the twentieth century will be to link up with studies of evolution being carried out today. Our journey will thus acquire a special interest for those who already know some biology. For the biologists at work today are pioneers on the same journey, cutting fresh trails on the same quest as earlier naturalists. The most modern research is not somehow isolated from that earlier work but is part of the progressive growth in our understanding of the living world.

It is often wise to consult a traveller's guide before setting out on a journey. The benefit and enjoyment of a journey can be much enhanced if we know beforehand what the main points of interest are and what to avoid. In the present case, we can do much the same thing by pausing to consider more closely the place that evolution occupies in modern science before passing to the seventeenth century. We shall be better prepared to embark once we understand clearly why evolution is such an exciting and important scientific concept.

CHAPTER ONE

EVOLUTION — JOURNEY'S GUIDE

Sherlock Holmes once remarked to Dr Watson that our ideas must be as broad as nature if they are to interpret nature. He made this remark with reference to some of Charles Darwin's ideas about evolution, and what a fitting comment it is! The theory of evolution is well calculated to broaden the ideas of any thoughtful person, for it seeks to unravel the thread that ties all living organisms together.

It does this by suggesting that all animals and plants living on Earth today are the modified descendants of quite different animals and plants, which existed at earlier times. Such modification is possible because of inherited differences that occur between one generation and the next in all organisms. Evolution takes place when these differences build up in one direction over successive generations. Changes of this kind have apparently been going on for millions of years, and as a rule the later organisms have tended to be more complex and diversified than earlier ones. So evolution encompasses the idea that the animals and plants we see around us today are all descended from just a few simple forms of life that first arose on Earth.

Evolution is a notable example of a far-reaching scientific theory. Concepts like this provide a means of understanding the natural world in which we live. Without their aid, we may gaze admiringly at some part of nature but we will not understand it. For instance, rooms in the stately homes of a bygone era are sometimes adorned by glass cabinets containing dead animals mounted for display. Often enough, these cabinets contain an assortment of birds and mammals, or else collected seashells and fossils mixed up together. In this situation, animals are treated as mere curiosities or objects pleasing to the eye, along with fine furniture and paintings. The mounted animals may be much admired, but some general theory is needed to explain them, if they are to be viewed with understanding.

An accurate drawing of the eastern grey kangaroo, from a late nineteenth-century book on mammals. This animal's scientific name, Macropus giganteus, *indicates the particular species to which it belongs (*giganteus*) and the genus of typical kangaroos (*Macropus*) in which this species is included.*

The Great Gray Kangaroo (*Macropus giganteus*).

Scientists do not adopt a particular theory just because it is an elegant idea. It must also provide a framework for solving difficulties that they encounter in their daily work, as they carry out research into the details of the natural world. Evolution has been adopted as a central theory in biology precisely because it enables us to solve a number of major puzzles that confront anyone who studies living things in any detail. The sorts of problem that crop up can be illustrated by any well-known group of animals or plants.

Take kangaroos, for example. Everybody knows that kangaroos live in Australia. Their form is now so familiar that a kangaroo motif adorns all manner of things Australian, from postcards to airliners. However, this very familiarity tends to make us forget what unusual animals these are. There is nothing quite like them anywhere else in the world. In fact, the first European naturalists to visit Australia were so astonished by the appearance of kangaroos that they had difficulty in drawing them correctly (see colour plate 1A).

The most distinctive feature of kangaroos is their powerful hindlimbs and long feet, which are used in a fast hopping gait. Their forelimbs are much smaller and do not touch the ground, except when the animals are at rest or moving slowly. No other large mammals use this kind of hopping movement. Another notable characteristic of kangaroos is that their young are born at an early stage, while they are still tiny, and are then nourished externally in a pouch. This pattern of reproduction is shared by a large group of mammals, the marsupials, which include koalas, possums and wombats. The marsupials are found mainly, but not exclusively, in Australia.

Kangaroos come in a variety of shapes and sizes. As a first step in studying such variety, biologists classify animals and plants into groups of similar forms. The basic unit of classification is the species. A species consists of individuals that resemble each other closely, except for differences due to age and sex, and differ consistently from individuals belonging to all other species. Species that are similar to one another are grouped together in higher categories, starting with the genus.

Taken together, typical kangaroos and their close relatives have been classified into more than fifty different species. These include a wide variety of animals with the same basic body plan, ranging in size from the small and distinctive rat-kangaroo to the large red kangaroo. There are also marked differences in their geographical distribution, with some species being widely distributed across the continent and others being confined to surprisingly small areas. But whether their geographical range is large or small, all these species are found on the Australian mainland or on nearby islands.

For anyone who wants to understand the living world, this information raises questions about the diversity of animals and their distribution across the world. Why, for example, are these species quite varied, and yet are all recognizable as species of kangaroo? Why do they show this pattern of unity amid the diversity? And why are they all confined to the Australian region?

Satisfactory answers to these questions are readily worked out in terms of evolution. For evolution makes sense of the fact that a large number of species, which share the same unusual body plan, are all found in the same part of the world. Such a pattern is to be expected if all these species are the modified descendants of some early marsupial mammals that had something like that basic body plan. That this whole range of species is confined to the Australian region is accounted for by the isolation of this island continent, which effectively inhibits migration by most mammals.

Another question here is: why do some kangaroos live in trees? Typical kangaroos are well adapted to bounding across the open plains, but a few species are at least partly adapted to living in trees. These are the tree kangaroos, which inhabit the tropical forests of New Guinea and the far north of Australia. The fact that each species of animal or plant is adapted to its way of life within a particular habitat raises as many questions as the diversity of species. Tree kangaroos illustrate this point beautifully.

Tree kangaroos are adapted to clambering about in trees by a number of obvious features (see colour plate 1B). They have forelimbs that are relatively longer and more muscular than those of typical kangaroos and these are used for hanging on to tree trunks and branches. All four feet bear long curved claws and have granular soles, which improve their grip. The tail is particularly long to aid in balancing. In addition, most species of tree kangaroo can move their hindlimbs alternately, enabling them to walk and run in trees rather than being limited to hopping, as are the ground-dwelling species.

So how come these animals are adapted to living in trees? The theory of evolution suggests that tree kangaroos are just what they seem to be: kangaroos that have become modified in certain features to fit them for life in the trees. That is to say, the process of evolutionary change is a process in which some pre-existing structures are modified to suit a new or changing environment. A strong indication that tree kangaroos really are descended from ground-dwelling kangaroos is provided by the way their feet are adapted to life in the trees.

One striking feature of the possums, which dwell in trees, is that the hind foot has a mobile big toe. This is offset from the other digits and so enables the animal to grasp the branch of a tree securely. In contrast, the ground-dwelling kangaroos lack the big toe entirely, and the next two digits are quite small. This loss or reduction of the lateral digits is an adaptation for speed over the ground, and is also found in other fast-moving mammals.

Now, in the tree kangaroos, the hind foot resembles that of other kangaroos in lacking a big toe and in having much reduced second and third digits. Useful though a grasping big toe would be, they do not

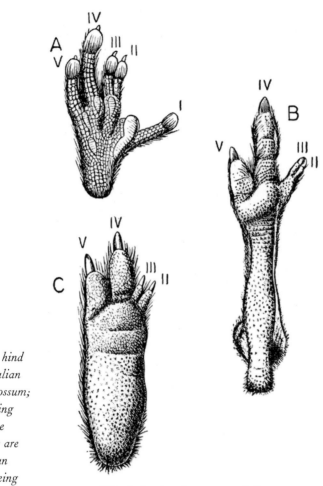

Adaptations of the hind feet in three Australian marsupials. A, a possum; B, a ground-dwelling kangaroo; C, a tree kangaroo. The toes are identified by Roman numerals, with I being the big toe.

Hind feet of marsupials.

possess one. Instead they obtain a grip on tree branches by means of the granular sole, which wraps round the side of the foot, and the curved claws. That this adaptation has been produced by modifying a pre-existing structure is clear because the adaptation is far from perfect. It has a makeshift quality about it, which makes it easy to see evolution at work.

This raises the further question of what kangaroos are doing in trees in the first place. On the island of Borneo, which is no great distance away and not very different in climate from New Guinea, there are apes and monkeys in the trees. In fact, from Borneo westwards, placental mammals dominate the scene, while east of Borneo marsupial mammals predominate. The absence of placental mammals, including apes and monkeys, from Australia and New Guinea is probably due to the isolation of these large islands. Consequently, there was an opportunity for living in the forests of tropical Australia and New Guinea, which could be exploited by the modified descendants of ground-dwelling kangaroos.

Whether Australia and New Guinea really have been separated from South East Asia for many millions of years as this theory requires is a matter of geology. It is the business of geology to reconstruct past changes in the structure of the planet by examining traces left in the rocks. Evidence derived from deep-sea drilling shows that the sea floor is continually spreading, with the result that the continents change their positions over geological time. From this, it appears that Australia and New Guinea, which lie on the same continental plate, were even more isolated in the past than they are now.

Kangaroos, then, are just one example of how puzzles about the adaptation and diversity of species can be resolved by the theory of evolution. If some early marsupials became isolated in Australia and subsequently evolved into a variety of species as opportunity arose, then this would explain why a whole range of kangaroos, including the remarkable tree kangaroos, are found only in Australia and its immediate neighbourhood. Furthermore, this conclusion dovetails well with the conclusions that geologists have reached concerning the movements of the continents in the remote past.

Of course, evolution would hardly be a matter of great interest if it applied only to some fifty species of kangaroo. However, it has been found to make sense of the situation time and again when animals and plants have been studied in detail. Many different branches of biology have combined to tell the same story, as the following chapters will show. It is therefore logical to regard evolution as a principle that applies generally to all living organisms. This being the case, the theory of evolution becomes one of the most important and illuminating contributions not only to biology but also to the whole of natural science.

In throwing light on the diversity and adaptations of living things, the theory of evolution can be seen to address two fundamental aspects of biology. This is reflected in the theory having two parts, though the distinction between them can never be absolute. One part of the theory deals with what has happened in the history of life on this planet, and the other part deals with how it has happened. The first part, then, is concerned

[*From the* JOURNAL *of the* PROCEEDINGS OF THE LINNEAN SOCIETY *for August* 1858.]

On the Tendency of Species to form Varieties; and on the Perpetuation of Varieties and Species by Natural Means of Selection. By CHARLES DARWIN, Esq., F.R.S., F.L.S., & F.G.S., and ALFRED WALLACE, Esq. Communicated by Sir CHARLES LYELL, F.R.S., F.L.S., and J. D. HOOKER, Esq., M.D., V.P.R.S., F.L.S., &c.

[Read July 1st, 1858.]

London, June 30th, 1858.

MY DEAR SIR,—The accompanying papers, which we have the honour of communicating to the Linnean Society, and which all relate to the same subject, viz. the Laws which affect the Production of Varieties, Races, and Species, contain the results of the investigations of two indefatigable naturalists, Mr. Charles Darwin and Mr. Alfred Wallace.

These gentlemen having, independently and unknown to one another, conceived the same very ingenious theory to account for the appearance and perpetuation of varieties and of specific forms on our planet, may both fairly claim the merit of being original thinkers in this important line of inquiry; but neither of them having published his views, though Mr. Darwin has for many years past been repeatedly urged by us to do so, and both authors having now unreservedly placed their papers in our hands, we think it would best promote the interests of science that a selection from them should be laid before the Linnean Society.

Title and opening sentences of the introduction to the Darwin/Wallace papers on species, published by the Linnean Society in 1858. This joint publication was the first to put forward natural selection as the mechanism of evolution.

with evolution as history. The theory seeks to explain the diversity of animals and plants by providing an accurate account of changes in the history of life that have given rise to this diversity. Inferences about what must have occurred in the past are based on a consistent pattern of evidence from various sources in the present.

The second part of the theory deals with evolution as a mechanism of change that produces well-adapted organisms. Here the theory seeks to explain how evolution occurs as a consequence of everyday processes in living things. Ideas about the mechanism of evolution are derived from evidence of processes at work in natural populations at the present time. The best known idea for explaining how adaptation is produced by evolutionary change is that of natural selection, first proposed by Charles Darwin and Alfred Wallace.

In noting that evolution concerns the history of life, we come to a point that distinguishes biology from chemistry and physics. For chemistry and physics, as these subjects are usually taught, deal with matters that are not affected by historical change. Apart from the origin of the universe and the formation of the chemical elements, they deal with structures and processes that are thought to be unchanged through time and space. The dramatic success of biochemistry and molecular biology shows that these physical structures and processes underlie everything that happens in a living organism. Hence the great complexity of living organisms does not, on its own, separate biology from chemistry and physics.

What makes the difference is that the complex structures studied in biology are the products of history. The example of kangaroos shows us how the present structure and distribution of living organisms can be explained by an historical sequence of events. Such a sequence of events produces species which are unique to their time and place and which might have been different or not existed at all if the flow of history had taken a different turn. Certainly, the unfolding history of life involves species in the past that were quite different from those living today.

It is sometimes possible to tell what these bygone animals and plants looked like because their remains have come down to us as fossils preserved in the rocks. In the study of these fossilized remains, biology joins forces with geology, which also has history as a central concern, and this joint study provides the most direct evidence available about how life has changed in the past. Probably the best known of all fossils are the dinosaurs, which dominated the Earth many millions of years ago but are now extinct. When dinosaurs and other extinct animals were dug up in ever-increasing numbers during the nineteenth century, they brought to light a previously unsuspected dimension of history.

The vanished worlds revealed by these fossils are now so familiar to us that it is hard to appreciate what an impact the early discoveries made. Some idea of their impact can be gained from the large sums people were willing to pay for the fossils. The British Museum, for example, paid £1300 to Albert Koch in 1844 for a fossil collection that included one of the best mastodon skeletons ever recovered. A few years later, in 1862–3, they parted with £700 to obtain another collection that included

the first specimen of the bird, *Archaeopteryx*, perhaps the most famous fossil of all.

These discoveries still have an impact on our minds when we reflect on how dramatic some of the changes have been. Imagine being able to travel back to the time of the dinosaurs, accompanied by three scientists: a chemist, a zoologist and a medical specialist. On stepping out of our time machine 100 million years ago, the chemist would find the same chemical elements undergoing the same interactions as they do now. He could continue his research as if nothing had changed, but the zoologist would find his world radically altered. Indeed, the zoologist might well fail to find a single species of animal that he recognized and his whole pro-gramme of research would have to be modified to suit the new conditions. Finally, the medical specialist would have no subject of research at all, and would have to be retrained as cook and bottle-washer for the other two.

Now this reference to the absence of humans touches upon a point that holds perhaps the greatest fascination of all. As a theory that aims to give an account of the history of life, evolution should include an account of human origins. For the history of life is our history, and this gives evolution an emotional impact that few scientific theories can match. This impact is profound because an accurate account of human origins must act as a constraint in philosophical debate about humanity's place in the scheme of things. And this weighty debate is something that thoughtful

The excitement caused by the discovery of spectacular extinct species is captured in this scene from a monograph published in 1799. The jaws of a gigantic fossil reptile are being recovered from the Chalk formations, underground. The fossil was later termed a mosasaur.

people in any age must reflect on at some time in their lives. It is a dull person indeed who can raise no interest in the matter.

Charles Darwin was perfectly well aware of the interest and importance of human evolution when he wrote the *Origin of Species*, but the emotional involvement was a complication that he wished to avoid. So he put the case for evolution without reference to human origins, contenting himself with a single sentence in the final chapter. 'Light will be thrown on the origin of man and his history' was all he said. Darwin's precautions were in vain. When the *Origin of Species* appeared in 1859, popular commentators promptly seized on this one sentence and dubbed his view the 'ape theory'.

Excitement over the possibility of our descent from ape-like ancestors was heightened by the fact that knowledge of the great apes was still quite new. By the early nineteenth century, naturalists had been able to give accurate descriptions of the chimpanzee from Africa and the orang-utan from Asia (see colour plate 2). But the larger African ape, the gorilla, was not described scientifically until the mid-nineteenth century. In 1861, the explorer, Paul du Chaillu, caused a sensation in

London when he lectured on his encounters with gorillas in the wild, and brought along a stuffed gorilla as an exhibit. Little wonder, then, that the gorilla featured in cartoons on evolution which appeared around that time.

In the years following publication of the *Origin of Species*, one of the things that gave the debates a good deal of zest was that evolution appeared to contradict the account of human origins found in the early chapters of Genesis. The book of Genesis, which is part of the Christian and Jewish Scriptures, teaches that the whole world, including animals, plants and humans, was created by God. It seemed to many that there was a conflict between creation and evolution, and some saw this as part of a more general conflict between science and religion as a whole. Since then, professional theologians have accommodated evolution within their thinking and the hubbub has died down. However, some Christian groups continue to reject evolution, and they have marketed the old conflict with new and glossier packaging in more recent times.

Faced with such acrimony, scholars are sometimes tempted to read the modern situation back into earlier times, and to portray a sharp conflict extending back for centuries into the past. In fact, such a view represents a serious historical error and it is imperative that we put it behind us in

A mid-nineteenth-century plate depicting a female gorilla and her offspring. This appeared in a scientific Memoir on the Gorilla *published by the anatomist, Richard Owen, in 1865.*

order to understand the narrative that follows. Before the nineteenth century, the pioneers of science simply did not see themselves as involved in a major conflict between science and religion, nor was there any great debate over evolution. On the contrary, science and religion were widely regarded as in harmony and mutually supportive. Indeed, in England much useful research in natural history was carried out by men who were themselves ordained ministers of the church.

Hence the way was prepared for evolution, not by men undertaking a dry run of late-nineteenth-century debates about science and religion, but by generations of naturalists who looked at the world quite differently. With the revival of learning in the sixteenth and seventeenth centuries, these naturalists could draw on three chief sources of ideas in their study of the living world.

First of all, there was the body of philosophy that had been inherited from the classical period by way of the medieval commentators. The most relevant part of this was the work of Aristotle, who had made extensive observations on a range of animals and had thought deeply about what he saw. He appreciated the diversity of life and came up with ways of differentiating the various natural groups of living things. Aristotle also recognized that the parts of an animal fit together to adapt it to the environment in which it lives, and he thought about the causes of this adaptation. This work and other contemporary knowledge of animals was summarized in Aristotle's famous zoological books, and these represent the first attempt to develop a science of natural history.

Secondly, there was the Christian religion, which provided the generally accepted framework of ideas in European society at that time. As part of this framework, the belief that the natural world had been created by God at the beginning was taken for granted. Science began to flourish in this context because the concept of a created world gave grounds for regarding nature as orderly and open to rational inquiry. The study of nature could be acknowledged and respected as the study of God's handiwork in creation.

So it came about that natural history was viewed as an appropriate activity for a clergyman in the Church of England, even if it was not part of his job. As late as 1831 it could be said that 'the pursuit of Natural History, though certainly not professional, is very suitable to a Clergyman'. This was the opinion of Josiah Wedgwood, who was Charles Darwin's uncle. And it helped persuade Darwin's father to let his son join the now famous expedition on board HMS *Beagle*, while still expecting him to take up a career in the church.

The third source of learning was the fledgling discipline of science, or the new philosophy as it was then known. In England, the new philosophy was championed by Francis Bacon (1561–1626) and his successors in the Royal Society of London. Bacon harmonized this new, scientific approach to the natural world with the theological framework of the day by acknowledging that both Nature and the Bible had their source in God. Therefore both these sources of illumination, the book of God's work and the book of God's word, deserve our closest attention. At a later date, a quotation from Bacon along these lines was to appear opposite the title page of the *Origin of Species* (see illustration on page 131).

A crucial innovation of the new philosophy, as applied to natural history, was that animals and plants began to be studied for their own sake, and statements about them began to be checked by first-hand observations and experiments. This approach sounds like plain common sense to us today, but back then it marked a significant departure from earlier traditions. Hitherto, the natural world had been regarded largely from a human-centred point of view. It was believed that animals, vegetables and minerals of all kinds had been designed and distributed by God with human needs in mind. Each species was therefore intended to serve some human purpose, which might be practical, moral or symbolic.

The chief focus of medieval literature about plants and animals was this rich assortment of moral associations and symbolic meanings. The medieval writers garnered these meanings from old and respected sources, from which they also took most of the natural lore about the creatures they were discussing. They were not concerned to record details of the natural world or to analyse its workings. Hence many a far-fetched tale was handed on because it could be traced back to an authority like Aristotle or one of the later compilers.

From the sixteenth to the eighteenth centuries, naturalists took the crucial step of gradually breaking away from these traditions and basing their accounts of animals and plants on their own work. Books on natural history were increasingly compiled from these recent observations, which enabled the writers to set aside many of the old fables as 'the errors of the ancients'. Just what a significant transition this was can be seen by comparing work from before and after this transition.

The need for first-hand observation in natural history is shown by this illustration of a fish, which was said to have the form of a monk. It comes from a book published by Pierre Belon in 1553.

For example, in an English *Bestiary* from the mid-thirteenth century, a couple of pages are devoted to the hyaena. Translated by R. Barber, the account begins: 'There is an animal called the hyaena, which lives in the graves of dead men and feeds on their bodies.' This cheerful statement is accompanied by an illustration of a hyaena taking its meal (see colour plate 3A). The text goes on to relate 'many marvellous things about this beast', some of which bear traces of someone having seen the animal while others are entirely fanciful. There is also a full discussion of the hyaena's symbolic significance, with a string of quotations from the Bible.

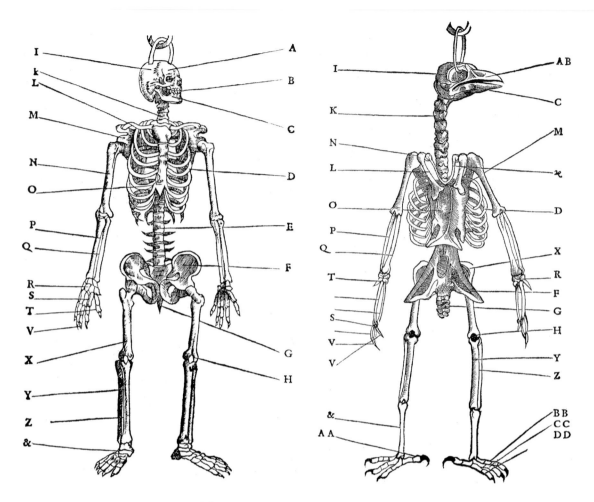

By contrast, a more matter-of-fact account of the hyaena's appearance and scavenging habits can be found in Oliver Goldsmith's popular *History of the Earth and Animated Nature*. First published in 1774, his account shows a real familiarity with the species in question, noting features such as the shape of the head and the distribution of hair over the body. These features, and the animal's characteristic stance due to its relatively short hind limbs, are all rendered accurately in a painting in Charles D'Orbigny's *Dictionary of Natural History*. Published in 1849, this illustration is clearly based on first-hand observation (see colour plate 3B). In both of these books, the main focus has shifted to documenting the natural world as accurately as possible.

Having a look for oneself soon began to make a difference in the work of the early naturalists. The transformation from hearsay to pioneering science could often be seen in the work of individual naturalists, as in

The benefits of first-hand observation are shown by this illustration from Pierre Belon's book on birds, published in 1555. Here he has made a careful comparison between the skeletons of a bird and a man, and has labelled equivalent bones with the same letter of the alphabet.

the case of the French doctor, Pierre Belon (1515–64). His extensive studies in natural history included an interest in fish, where the influence of hearsay is evident in his account of a fish that had the form of a monk and was able to moan. However, Belon also studied birds, and here, among other things, he compared a human skeleton with that of a bird. His detailed observations went so far as to identify, correctly, the corresponding bones in each skeleton. This fine study in comparative anatomy is justly famous as a pioneering piece of science.

Our narrative, then, begins with naturalists in the latter half of the seventeenth century who drew inspiration from Aristotle, from the Christian context of the day and from the new philosophy. These people laid the foundations of modern biology by collecting plants and animals, and listing and describing what they found. One of the most important contributions to this effort was made by John Ray (1627–1705) in England. By looking at the work of Ray and his contemporaries in England, we can see how the success of their efforts brought them up against major puzzles about living organisms that would subsequently be addressed by the theory of evolution. Even at that early period, it soon became apparent that successful natural history must eventually deal with such far-reaching issues as the history of life and the status of the human species.

So let us turn to the work of the man described by another naturalist as 'the excellent Mr Ray'.

CHAPTER TWO

PUZZLES FOR THE NATURALIST

The head of King Charles I was still firmly set upon his shoulders when John Ray went up to Cambridge in 1644. The university continued to function in these turbulent times, but even the most gifted individuals might be dismissed if they spoke out on matters pertinent to the English Civil War. Ray came to the university from the village of Black Notley near Braintree in Essex, where his father was a smith. His obvious ability had gained him a place as an undergraduate at Trinity College. There, Ray received the training of a clergyman and, having genuine Christian convictions, he was later ordained.

Science had no place in the curriculum in those days. So it is probable that Ray took up the intensive study of science, and in particular botany, only after he had gained a graduate fellowship. Even then, he could find no one in the university to teach him how to identify plants. So he formed his own group of enthusiastic students to collect and sort out local plants. Ray's first book emerged from the activities of this group in 1660, a *Catalogue of Cambridge Plants*, which described 626 local plants.

Among this group of students, many were the sons of country gentlemen. One of these was Francis Willughby, who became a close friend of Ray in the study of natural history. Being a wealthy squire, Willughby was able to finance their joint projects when Ray lost his job for refusing to sign an oath required of all clergy by Charles II.

The central problem that Ray and his fellow naturalists faced was one of diversity. A steadily increasing number of species, especially of plants, were described across Europe as a result of their efforts. On top of this, excitement over the diversity of life was fuelled by the work of explorers, who discovered that other parts of the world were swarming with hitherto unknown forms of life. As if this were not enough, the invention of the microscope revealed the existence of an unsuspected multitude of tiny

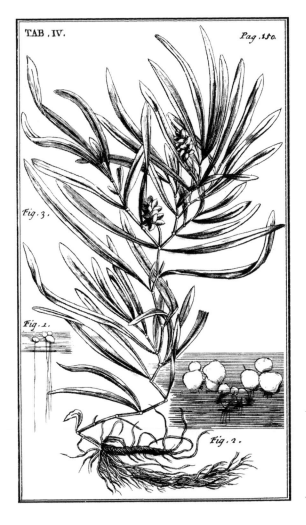

One of the common pondweeds belonging to the genus Potamogeton. *From Ray's* Synopsis *of the British flora, which first appeared in 1690.*

organisms. It soon became obvious that there was an enormous variety of animals and plants, and that it was by no means easy to distinguish them all correctly.

The foremost need of the period was therefore a sound classification. A classification is necessary wherever one has to deal with a diverse range of objects, whether these are books in a library, products to be marketed or animals and plants in nature. The process of classifying involves the grouping of individual objects into appropriate classes or categories. It is important to get a classification right, because, once established, it affects our whole way of looking at things.

Hence the important step taken by naturalists in the late seventeenth century was that they began to develop a new, more objective, classification and with it a new and more objective way of looking at the living world. Previously, animals and plants had been grouped largely in terms of their relationship to humanity, in so far as they had been classified at all. The emphasis had been mainly on their practical usefulness or on their moral and symbolic significance. This human-centred approach was gradually rejected by the naturalists, who tried increasingly to group living organisms by their structural features rather than by their usefulness or moral status.

A good example of the way animals were studied in the decades before John Ray began his work is provided by Edward Topsell (1572–1638), who produced the first illustrated book of natural history to be published in English. The interest in natural history was there in Topsell's book but it was swamped by the wide range of other interests. These are reflected in the book's full title, which is *The History of Four-footed Beasts and Serpents, describing at large their true and lively Figure, their several Names, Conditions, Kinds, Virtues (both Natural and Medicinal), Countries of their Breed, their Love and Hatred for Mankind, and the Wonderful Work of God in their Creation, Preservation and Destruction.* And this is followed by a subtitle of equal length!

The first edition of *The History of Four-footed Beasts* appeared in 1607. It drew heavily on the work of other naturalists and also on the older compilers. Consequently it repeated much of the old folklore, but it also included material from Topsell's own studies. The animals were classified to some extent, as the title shows, but the main focus was still the animals' relation to humanity, including their edibility.

In the study of plants, seventeenth-century publishing was dominated by the herbals, which dealt with plants in terms of their practical use in medicine and classified them accordingly. One English herbal published in the mid-seventeenth century was written by Nicholas Culpeper (1616–54), who described many common plants quite accurately but explained their uses in terms of astrology and alchemy. He was content simply to list, and illustrate, plants in alphabetical order (see colour plate 4). This left readers to identify a plant by thumbing through the book until they encountered something reasonably similar. But such a method could not cope with the thousands of plant species known to naturalists by the late seventeenth century.

In their efforts to forge a new classification that could cope with the expanding catalogue of animal and plant species, naturalists were able to

A fine woodcut of a bison from Topsell's History of Four-footed Beasts. *The relative accuracy of this illustration clearly reflects first-hand observation, as does Topsell's comment that 'through the admirable strength of his neck' the bison can toss a horse and its rider into the air.*

draw on the foundation provided by Aristotle many centuries before. As a philosopher, Aristotle had employed logical division as a method of defining entities in the world. This method involved a sequence of general classes, each subdivided into two or more special cases as judged by some essential differences. Thus all conceivable things might be divided into two classes, living and non-living. In turn, living things might be divided into animals and plants; animals might be divided into aquatic, terrestrial and flying animals, and so on.

In this system of logic, the class being divided at each level is termed the 'genus' and the two alternative subdivisions are the 'species', which in turn become the genera (plural of genus) for the next level of the subdivision. Hence this method provided two fundamental terms that are still used in biology today. Although the method of logical division provided the terminology and the conceptual framework, Aristotle did not use it to draw up a formal classification of living things.

Instead, he adopted a down-to-earth approach in his book on the *History of Animals*. By comparing a large number of animals, he was able to begin arranging individual species in to collective groups, the genera, by the 'resemblance of the shapes of their parts, or their whole body'. By modern standards, Aristotle's genera were rather large groupings; for instance, he recognized birds, fishes and cetaceans as separate genera, each containing many related species. The important point is that he established these groups by overall resemblance based on the comparison of many points of detail. In other words, his groupings were decided by inspection rather than by logic.

Aristotle himself never drew up an overall table of classification and he seems to have had little interest in classification as such. He did, however, rank his groups in some sort of sequence using a couple of different criteria. In his book on the *Generation of Animals*, for instance, he used the relative perfection of animals at their moment of birth. The result was a scale of perfection or ladder of nature running from man at one end through animals and plants to minerals at the other. Such is the complexity and variety of life that Aristotle could not always be sure just where a given group belonged. Thus the sea-anemones and sea-squirts were doubtful groups that seemed to be intermediate between plants and animals. Consequently his concept of a graduated ladder of nature carried with it the idea of continuity rather than sharply distinguished grades.

Naturalists of the seventeenth century built on Aristotle's approach to classification by using logical division as a method and by being quite flexible in its application. Their first-hand knowledge led these workers to appreciate the naturalness of certain groups both in animals and, less obviously, in plants. Somehow they managed to adjust their choice and sequence of divisions so as to avoid breaking up these groups. Hence a situation developed where many natural groups were evidently recognized by inspection, and these were then fitted into a classification by logical division for the sake of convenient identification. John Ray was one of those who used the principles of logical division in classification but did not hesitate to bend the rules in order to produce groups that he recognized as natural.

Together with his young friend, Francis Willughby, Ray had hoped to produce a soundly based classification of both plants and animals. The two men resolved to 'reduce the several tribes of things to a method and to give accurate descriptions of the several species' based on first-hand knowledge. They travelled widely across England and the Continent to collect material for this project. The plants were to have been mainly Ray's responsibility and the animals were to have been Willughby's, but the untimely death of Willughby prevented the execution of their plan. However, Ray went on to publish some of their joint material, of which their *Ornithology* is a fine example. First published in Latin in 1676, the English edition followed in 1678.

This book represented a major advance in animal classification, and one of the main reasons for this is that they chose to group birds according to their intrinsic structural features. They concentrated on the form of the beak, the structure of the foot and the overall size of the body. In this way, they were able to group together, in classes or genera, birds that truly resembled each other. Willughby and Ray explicitly rejected the human-centred, symbolic tradition in their *Ornithology*. At the outset, they stressed 'that we have wholly omitted what we find in other authors concerning . . . emblems, morals, fables, presages or aught else appertaining to divinity, ethics, grammar or any sort of human learning'. And they assured the reader that the book's treatment of birds will 'present him only with what properly relates to their natural history'.

In the *Ornithology*, birds were classified by logical division, being divided first of all into land and water birds; land birds were divided into those with curved beaks and straight beaks; birds with curved beaks were divided into birds of prey and fruit-eaters, and so on. This method was used in a flexible manner and kept up only until a compact final group was reached. In the case of large flightless birds, the final grouping was reached after only three subdivisions but more complex groups required up to eight subdivisions. Many of these final groups correspond well with groupings still recognized as natural today, such as crows (today the Corvinae), poultry (Gallinae), woodpeckers (Picinae), geese (Anserinae) and others.

Within each of these groups, the individual species were listed and described one at a time. Each species was given an English name and usually, but not always, a Latin name consisting of two words, one for the genus and the other for the species. Most of the species recognized by Willughby and Ray are well-known birds, which are still regarded as true species today.

While studying birds, Ray did not experience any great difficulty in distinguishing one species from another, but this was certainly not true for his study of plants. At that period, it was generally accepted that species represent different types of organism. Each one consists of similar individuals that are different from individuals belonging to other species. But as soon as one tries to classify living organisms, it becomes clear that they are highly variable. Plants especially tend to vary in the most irritating manner from one locality to another or according to the type of soil. Ray was therefore driven to think hard about how species,

Three species of owl, from the Ornithology *by Willughby and Ray, who had no difficulty in recognizing owls as a natural group of birds. The* Ornithology *included European as well as British species.*

as the basic unit of classification, could be distinguished from one another with certainty.

His answer was given in 1686. 'After a long and considerable investigation,' he wrote in the *General History of Plants*, 'no surer criterion for determining species has occurred to me than the distinguishing features that perpetuate themselves in propagation from seed.' It was evident to Ray that the range of variation that any given pair of parents can produce in their offspring must be contained within the potential of a single species. This applies as well to animals as to plants. So Ray used the biological criterion of reproduction for defining species as real entities and distinguishing them from varieties produced by different local conditions. Ray's definition of species was adopted enthusiastically by other naturalists and remained in use, with only slight modifications, for the next 150 years.

Faced with the problem of biological diversity, then, Ray and other naturalists responded by trying to devise a much better classification of living things. By the end of the seventeenth century, the main elements of the problem had become clear. It was obvious that there are many different species, each one consisting of similar individuals. The individuals within a species might vary to a disconcerting extent but the members of a species are all united by reproduction. Inspection revealed that several species may share a number of common features and so form a group of species that resemble each other. Ray, especially, was insistent that classification ought to reflect these natural groupings. This could not be done with an artificial system of logical division, however elegant and convenient this might be. Finally, the usefulness or symbolic value of species came to be regarded as quite irrelevant. The new classifications assessed animals and plants by their intrinsic features.

Not only are there a great many species, but also each one is intricately adapted to its way of life. The parts of an animal or plant are constructed so as to serve some function in the way of life followed by that species. This adaptation is such an outstanding characteristic of life that it strikes anyone who stops to think about the natural world. Certainly, it was clear to Aristotle, who saw that it demanded a satisfactory explanation. This he endeavoured to provide in his book on the *Parts of Animals*. The way Aristotle sought to explain adaptation was elegant enough, but by the late seventeenth century it had generated a painful paradox that was not resolved until the *Origin of Species* was published.

Aristotle spelt out his understanding of biological adaptation in terms of a comparison between animal organs and human artefacts. A human instrument such as a saw is made for sawing, and similarly the parts of animals exist for the sake of the function to which they are adapted. 'For just as human creations are the products of art,' Aristotle explains, 'so living objects are manifestly the products of an analogous cause or principle, not external but internal, derived like the hot and the cold from the environing universe.' Just as the function of a human instrument determines the structure of that instrument, so the function of a particular bodily organ actually exerts a causal influence on that organ. Aristotle termed this the 'final cause' of the organ's existence, and he considered this to be one natural cause among others.

If the parts of the adult animal are to be intricately adapted to their function in each generation, they must develop correctly in the embryo. Aristotle knew that the adult structure is produced by an elaborate sequence of changes in the embryo from his own study of the development of the chick within the egg. He recognized that each event in this sequence is triggered by the antecedent event in the sequence. Such a chain of events are the 'efficient cause' of the existence of the organism and its parts. At the same time, Aristotle considered that there must be a form-giving principle that controls development with the adult function in view. This corresponds, on the above analogy, to the blueprint in relation to the human instrument. He called this the 'formal cause' of the organism's existence.

This system of causes was designed to explain both the adaptations of the adult and their correct development in natural terms. Different though his world is from ours, it is not hard for the modern biologist to agree that Aristotle recognized the importance of adaptation and development, and attempted to grapple with the problems they raise in a sensible manner.

However, in common with other men of his time, Aristotle made no clear distinction between living and non-living things. The same questions, and the same causal answers, could be applied to anything involved in regular natural changes. The question, what is the function or purpose of it (final cause), could be applied to a stone as much as to the limb of an animal. The movement of a stone was also explained in a manner reminiscent of the explanation of animal development. Consequently, the physics of Aristotle is so different from ours today that the modern physicist finds it virtually unintelligible.

In the interval of many centuries between classical times and the rise of modern science, Christian ideas had come to dominate Europe. Aristotle's work was still treated with great respect and his understanding of nature was accommodated within this new conceptual framework. The genera and species of living things were now viewed as having been created by God in the beginning and as being conserved by Him in their original condition, rather than being eternal as Aristotle had supposed. And the adaptation of structure to function, which is so evident in the parts of animals, was now interpreted as due to divine design imposed on species when they were created. On this interpretation, final cause ceased to be internal and natural, and became external and supernatural.

When modern physics began to get under way in the seventeenth century, it involved a complete rejection of all Aristotle's causes except the efficient cause. Enquiries about the function or purpose of physical objects or events were banished since they led to no useful experiments. From then on, only the causal action of antecedent events would be recognized as scientific. Francis Bacon expressed this rejection by comparing final causes to the vestal virgins who served at Roman temples. Like them, he said, 'they are dedicated to God and are barren'. Later on, the French philosopher Descartes argued for the exclusion of all questions of final cause from science on the grounds that they have no use in physics or natural matters. Further, it is arrogant to think that we can find out God's intentions.

This rejection of Aristotle's approach made an awkward situation for contemporary naturalists, who were trying to study animals and plants scientifically. Questions about the purpose for which a stone falls to the ground might have no use in physics, but questions of design or purpose were clearly useful for understanding the structure and behaviour of living things. Therefore, the naturalists of the day were not happy to be rid of final causes. John Ray, in particular, argued strongly that final cause or design cannot be banished when considering the adaptations of parts of animals to their functions. His argument was set out in *The Wisdom of God Manifested in the Works of Creation*, first published in 1691. In this book, he used the adaptations of living organisms as evidence of the Creator's handiwork, based on his wealth of practical knowledge of animals and plants.

In *The Wisdom of God*, Ray contradicted Descartes' rejection of final causes. One of his main examples, the eye, was to remain a famous

Picus viridis
The common green Woodpecker or Woodspite.

Caput Pici dissectum

Picus varius minor
The lesser spotted Woodpecker.

Picus varius major
The greater spotted Woodpecker.

Three species of woodpecker, from Willughby and Ray's Ornithology. As well as showing the diversity of species, this figure shows the distinctive adaptations of woodpeckers, such as beak, toes and tail feathers.

one for a long time. The eye is employed by men and animals for vision, argued Ray, and is so admirably adapted to this use that all the wit of men and angels could not have done a better job. Therefore, 'it must needs be highly absurd and unreasonable to affirm, either that it was not designed at all for this use, or that it is impossible for man to know whether it was or not'. A telling point, this. It surely makes sense of the eye to regard it as an instrument designed for the purpose of seeing.

The same argument was also made effectively by cases in which several independent features are modified to produce a co-ordinated adaptation to some specialized way of life. Here, Ray was able to make good use of the ornithological material that he had gathered with Willughby. The specialization of woodpeckers for feeding on insects while clinging to tree trunks provides a perfect case in point. Ray noted the strong beak and long tongue, which are so well suited to obtaining insects from under the bark of trees. Then he drew attention to the adaptations for climbing trees, such as the short but strong legs, and their toes, which are disposed with two pointing forwards and two backwards. This disposition, commented Ray, 'nature, or rather the wisdom of the Creator, hath granted to woodpeckers, because it is very convenient for the climbing of trees, to which also conduces the stiffness of the feathers of their tails, and their bending downward, whereby they are fitted to serve as a prop for them to lean upon, and bear up their bodies'.

Such fine examples could not but command attention. There is no doubt that the emphasis on design in animals and plants encouraged the study of natural history. People were led to examine living things more closely, and they expected to uncover examples of adaptive mechanism as they did so. This is particularly true of the whole new realm of tiny structures that were being revealed by studies with that newly invented instrument, the microscope. One of the pioneers of microscopy was Robert Hooke (1635–1703), who was a leading member of the Royal Society in London. With his own microscope, he examined a variety of objects, including many living organisms or their parts. His observations were reported at meetings of the Royal Society, and later made available in his book *Micrographia*, published in 1665.

This was the first great work devoted to microscopy and the first to be accompanied by a full set of illustrations. Insects, which are so rewarding to the microscopist, were the subject of some of Hooke's most detailed descriptions and finest plates. These horrendous portraits of fleas, flies and lice, some over 30 cm across, must have truly astonished his contemporaries. In his diary for 1665, Samuel Pepys records that he sat up until two in the morning reading the *Micrographia*, which he thought was the 'most ingenious book' he had ever read.

In the case of fleas, Hooke not only admired 'the strength and beauty of this small creature' but also, with his keen engineering insight, explained how their limbs are adapted for jumping. Whether it was the legs of fleas, the feet of flies, or the barbs that keep a bird's feathers together, Hooke was impressed by the intricacy of adaptation revealed by his microscope. All this he saw as evidence of the Creator's design. Could anyone be so crazy, he wondered, 'as to think all these things the productions of chance?' Either that person's reasoning must be sadly amiss or else he has not attended properly to the works of the Almighty.

So there the matter stood. The naturalists clearly recognized biological adaptation as a most important fact of life. They took this close fit between

The house fly as seen under the microscope; a plate from Hooke's Micrographia. *Such fine drawings and the accompanying descriptions helped to extend the notion of adaptation and design to the very small.*

an organism's structure and its way of life to be a product of the Creator's design, imposed at the moment when each species was first created. This view had the benefit of promoting a keen interest in the adaptations of animals and plants. It also forged a positive link between science and the Christian ideas that provided the framework of contemporary society. But it suffered from the disadvantage of allowing that adaptation could not be explained by natural causes. Design and final cause became part of theology and were not part of science as they had been for Aristotle. In the long run, this was to cause no end of trouble.

Hooke's *Micrographia* also raised another matter that was to give naturalists a good deal of trouble. One of the objects that he examined was a piece of fossilized wood, and he used his microscopical observations to introduce a brief general essay on the origin of fossils. These remarks in the *Micrographia* seem quite ordinary to the modern reader but this is because we take it so much for granted that fossils are the remains of ancient, and often vanished, forms of life. To those who first read them, Hooke's remarks seemed bold and unorthodox. For he wrote at a time when no one had any idea that there had been great changes in the history of life and when the world was thought to be only a few thousand years old. The evidence of fossils began to change all that.

The situation in Hooke's day came about in this way. Looking back to the classical period, there was no foundation for the study of past time comparable to that laid for natural history. Plato and Aristotle were searching for general principles that would explain the flow of events that we experience. Beneath this flow of temporal changes, they looked for an order of nature that was eternal and unchanging. With this perspective, any historical sequence of events, whether in nature or society, was not considered to be of any great importance. So the success of the Greek philosophers in breaking new ground in so many areas was closely linked with a rather cavalier attitude toward the past.

The rise of the Christian religion transformed the classical view of the world. Christianity inherited from Judaism a strongly directional concept of world history. Both of these religions depend on the recorded actions of God at particular moments in their past, such as the exodus, the giving of the law, or the birth of Christ. From this perspective, the course of historical events acquires a profound significance and becomes a natural subject for scholarly enquiry. Drawing on the Bible, theologians of the early centuries developed a popular Christian cosmology, which ran from the creation and flood in the past to the second coming of Christ and the millennium in the future. At a more scholarly level, accurate dating of the past was encouraged, and a fine example from the fourth century is the *Chronology* of Eusebius, which provided the foundation for the dating system of the Western world.

With the revival of learning in the sixteenth and seventeenth centuries, a rigorous attempt was made to carry dating back even to the beginning of the world. The Old Testament narrative in the Bible was used as a source of information about the remote past along with other ancient manuscripts. Within this context, dating of the past was improved with the aid of Kepler's cycles of solar eclipses and other new sources. Chronology was accepted as a study of the highest respectability and was taken up by many able scholars. When Archbishop James Ussher published his *Annals of the Old Testament* in 1650, with its now notorious date of 4004 BCE for the creation, it was not frowned upon as a piece of religious dogmatism but, on the contrary, was welcomed as a fine example of contemporary scholarship. Ussher's estimate was perfectly reasonable, given the evidence available to him in the seventeenth century.

The first category of evidence about the past, which would eventually shatter this chronology, was that of fossil remains embedded in the rocks. Down the centuries, people's attention had often been caught by fossil bones and teeth that were exceptionally large. These were frequently interpreted as the remains of human giants, who had lived in ancient times according to many a legend, and so their true significance was overlooked. One famous example was described and illustrated in 1677 by Robert Plot (1640–96), a clergyman-naturalist with a position at Oxford University. By examining its structure, Plot recognized his find as 'a real Bone, now petrified', and correctly identified it as the lower end of a thigh bone or femur.

From its sheer size, he concluded that the bone must belong to an animal larger than an ox or horse, and he considered the possibility of an elephant brought to England by the Romans. However, when Plot had the chance to examine the skeleton of an elephant, its femur proved to be different from his fossil and so he fell back on the idea of a human giant. From his detailed account, we can now tell that this was the femur of a large carnivorous dinosaur belonging to the genus *Megalosaurus*. In fact, this is often celebrated as the first illustrated description of a dinosaur bone. But no such interpretation was possible in Plot's day, and the human giant idea gave it the air of a curiosity rather than an important discovery.

Another reason why it took a long time for the significance of fossils to be appreciated was that the designation 'fossil objects' originally included a much wider range of things than just the petrified remains of animals and plants. It was recognized that some fossils resembled living organisms but others appeared to resemble geometrical figures or heavenly bodies. Many of these resemblances were explained away as 'sports of nature' formed within the rocks by some strange 'plastic virtue' inherent in the Earth. Thus Robert Plot, among others, considered that this was the true explanation for the presence of numerous fossil shells in the rocks: they were ornaments of nature and not organic remains. Eventually, the key question

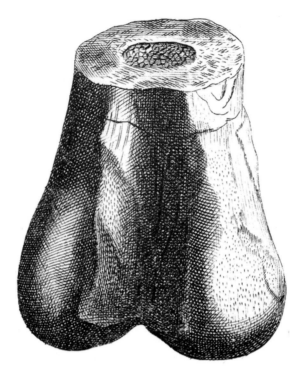

Robert Plot's illustration of a fossil bone dug out of a quarry near the town of Chipping Norton in Oxfordshire. Plot recognized the fossil as the lower end of an enormous femur (thigh bone), but some more imaginative writers interpreted it as a giant scrotum.

was seen to be: how could one establish whether or not a stone object that resembled an animal or plant really was of organic origin?

This was the question that Hooke tackled in his *Micrographia*. He expanded the topic in his lectures to the Royal Society but these were not printed until 1705, after Hooke's death. He observed that the microscopic structure of fossil wood closely resembled that of charcoal or rotten wood and naturally concluded that the fossil was also derived from a once-living tree. He supposed that the stony nature of the fossil was due to its 'having lain in some place where it was well soaked with petrifying water', which had impregnated it with 'stony and earthy particles'. Petrifications could also come about, Hooke realized, through impressions or moulds left by the original organisms on the surrounding materials of a suitable consistency, 'such as heated wax affords to the seal'. In addition, casts might be formed by materials introduced into these natural moulds.

Having thought out how petrifaction could occur, Hooke was left in no doubt that fossils were the petrified remains of living organisms. As well as fossil wood, he paid special attention to a group of fossil shells known as the ammonites. These had received their name, *ammonis cornu*

(Latin for ram's horns), on account of their coiled shape, more than a century before. However, some of the more loosely coiled shells were classed as 'serpents stones', from the popular notion of their resemblance to a coiled snake. With his insight into petrifaction, Hooke was able to sort out the confusingly preserved ammonites and recognize them as a single group of shells. More than that, he cut longitudinal sections through them and so established their similarity to the chambered shells of the pearly *Nautilus*, which had only recently been described from explorations in the East Indies.

In accepting that ammonites were actually the relics of once-living organisms, Hooke was faced with the problem of these marine shells being found inland and even in mountains. The discussion in his *Micrographia* casually attributes their occurrence on land to 'some deluge, inundation, earthquake, or some such other means'. He considered the point more carefully in his lectures and realized that the surface of the Earth must have been transformed since the creation. 'Parts which have been sea are now land,' he explained, 'mountains have been turned into plains, and plains into mountains, and the like.'

It was Neils Stensen (1638–86) in Italy who considered these consequences more fully. Stensen, more commonly known as Steno, had received medical training and become a renowned anatomist. He had also met Hooke when on a visit to London. Steno's interest in fossils was roused when he examined the head of a large shark at the request of his patron, Duke Ferdinand II. He saw that the shark's teeth strongly resembled the fossil objects commonly known as 'tongue-stones', and he concluded that tongue-stones were in fact fossilized shark's teeth.

In 1669, Steno published the *Prodromus* or forerunner of an intended *Dissertation* on the subject of fossils. Though the *Dissertation* was never written, the *Prodromus* on its own was enough to revolutionize the subject. Steno considered that fossils of organic origin must differ in their mode of growth from inorganic 'fossils' that did 'grow' within the earth, such as rock crystal. He analysed quartz and pyrite crystals and concluded that they grow by accretion of particles precipitated from solution just like crystals in the laboratory. The shells of molluscs, by contrast, show a different pattern, in which particles accumulate only along their outer edges, reflecting the growth pattern of their occupants. Fossil shells resemble living shells in this respect and so their organic origin is confirmed.

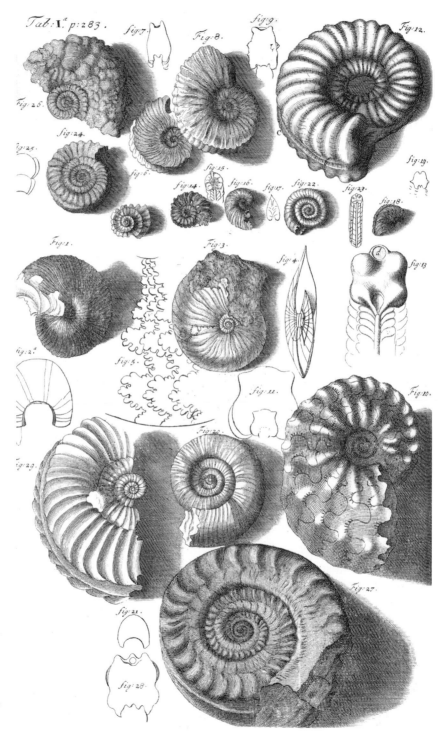

Ammonites drawn by Hooke, printed in his Lectures and Discourses of Earthquakes *in 1705. The little sketches beside the main drawings show that Hooke correctly understood the construction of ammonite shells.*

Then Steno considered the relative times of hardening of the fossil and the rock that enclosed it. As in Hooke's sealing wax analogy, it was evident that the fossil shells must have been hard while the rock was still soft because it was the fossils that had left their impressions on the rocks. This led Steno to reinforce the conclusion, formed from his study of tongue-stones, that the rocks had originally formed as soft muddy sediments in water. The layered structure of the fossil-bearing rocks in Tuscany implied that sedimentation had occurred in discrete episodes and the rock layers or strata had not been formed all at once. When any given layer was being deposited in lake or sea, the material above it must have been fluid and 'therefore, at the time when the lowest stratum was being formed, none of the upper strata existed'. Thus Steno made a most important inference, that a stack of solid strata in sedimentary rock implies a sequence of events in past time. In undisturbed strata, the lowest was deposited first and hence is the oldest.

To this Steno added another important inference. He pointed out that particles settling at the bottom of a body of water tend to accumulate horizontally. Hence rocks formed by sedimentation must all have been formed originally with the strata disposed horizontally, and the tilted position in which such rocks are commonly found today must reflect changes subsequent to their original deposition. By means of these inferences, the present arrangement of sedimentary rocks in a district could be used to reconstruct a sequence of events in the history of the Earth.

Between them, Hooke and Steno carried their arguments remarkably far. Once persuaded that fossil objects resembling animals really are the relics of animals that once lived, they could recognize fossils as being evidence about the past. Before this, their importance was not appreciated. Having correctly understood the nature of fossils, Hooke and Steno saw them as a key that would open a new interpretation of world history. As Hooke explained, we employ Roman coins or urns as clues to interpret a past civilization, and in the same way fossils are 'the medals, urns or monuments of nature.'

Important though this understanding of fossils was, it did not immediately lead to a lengthening of the geological timescale. Although Hooke wrote about geological upheavals in a way that would seem to imply an earlier date for creation than was commonly accepted, his published comments on the extent of time were always ambivalent. Steno, for his part, had no problem with the short timescale. His insights about fossils and strata were all accommodated within the 6000 years or so that was commonly accepted for the age of the Earth. Such an accommodation was possible because he took it for granted that the main geological event in Earth's history had been the Biblical flood. This assumption was shared by others who began to think about a scientific approach to global history in the late seventeenth century.

One of Hooke's favourite books was the *Sacred Theory of The Earth* by Thomas Burnet (1635–1715). The first part of the book appeared in Latin in 1680 and the complete work was available in English by 1690. A man of wide learning, Burnet set out to amplify the Biblical outline of world history with the latest available scientific thinking. This speculative approach took its cue from Descartes, whose *Principles of Philosophy* had included a hypothetical outline of planetary history based solely on the matter and motion of the parts of the Earth. Descartes imagined a hot, star-like origin, followed by cooling, which would lead to the formation of an outer crust that would crack and collapse as cooling continued. In some such manner, he supposed, the natural features of this or any other world could be accounted for by the new philosophy.

The flood recorded in the Bible was the pivotal event in the first part of Burnet's book. Before the flood, he thought the world had been smooth and spherical, without seas or mountains, and with a climate of perpetual spring. But then the waters burst upon a sinful humanity and devastated the surface of the entire globe, a scene that inspires a corresponding turbulence in Burnet's pen. The Earth that we now inhabit is but the ruin of a former world. Burnet proceeded to explain this sequence of events after the manner of Descartes, by reference to causes inherent in the construction of the planet. The flood was caused by the heat of the Sun drying out the crust of the Earth so that it cracked and collapsed. This, in turn, caused the waters below (the 'great deep' of the Bible) to boil up and burst over the surface of the planet. All this was accomplished by natural causes. Concerning those who employ God to create, and afterwards to annihilate, miraculous quantities of water for the flood, Burnet remarked that 'methinks they make very bold with Deity'.

With its controversial thesis and delightful prose, it is hardly surprising that the *Sacred Theory of the Earth* attracted much attention and was highly influential. For instance, Isaac Newton (1642–1727), who brought physics to a remarkable peak of achievement with his *Principia*, was publicly cautious about such speculative essays. However, in private he wrote

Frontispiece of Burnet's Sacred Theory of the Earth. *Christ stands astride the first and last stages of the Earth's development, shown in a clockwise sequence of seven spheres. Burnet tried to explain this sequence of events in terms of a succession of natural causes.*

enthusiastically to Burnet, saying that 'of our present sea, rocks mountains etc. I think you have given the most plausible account'. Again, after making some alternative suggestions, he adds: 'All this I write not to oppose you, for I think the main part of your hypothesis as probable as what I have here written, if not in some respects more probable.' As the example of Newton shows, Burnet was seen in his own time as speaking in favour of a reasoned understanding of world history.

The overall effect of Burnet's literary skill was to help create a climate of opinion, in which people would seek an account of the whole history of the Earth in terms of a succession of natural processes. This was all very well but it soon became apparent that his theory failed to answer many problems. In particular, it had ignored the evidence of fossils. Burnet's theory actually made it more difficult to account for the presence of marine shells high up in mountains since there were neither seas nor mountains before the flood in his scheme.

This was a deficiency that a physician by the name of John Woodward (1665–1728) sought to make good in his *Essay Toward a Natural History of the Earth*, published in 1695. Woodward had the advantage that, unlike Burnet, he had some practical knowledge of geology. He proposed that the crust of the Earth did not collapse in ruins in the flood, but rather was 'dissolved' and churned up by the floodwaters. Then the particles settled out in order of their specific gravity to form the rock strata with their entombed fossils. On this view, following Steno, the fossil-bearing strata had been laid down horizontally at the time of the flood and the fossilized organisms dated from before the flood.

All these developments were followed closely by the judicious John Ray. He felt that neither Burnet's nor Woodward's theories accounted for the presence of marine fossils in mountain rocks. In the case of Woodward's specific gravity theory, this failed a direct test, for it simply was not true that the largest fossils were always found in the lower strata. But then Ray remained irresolute on the question of the organic origin of fossils until the end of his life. If fossils like ammonites were truly organic in origin, then that implied that they were extinct, unless perhaps such creatures still lurked undiscovered in some recesses of the ocean. And for Ray and many others, extinction seemed to be incompatible with the providence of God.

On top of this difficulty, any theory based on the flood might have to be abandoned to account for the occurrence of marine fossils in mountains. Instead, it might be necessary to allow for successive rising and falling of the land, as Hooke had suggested. That in turn would imply that the world must be much older than current chronologies allowed. However, Ray's faint glimpse of an account of fossils without the flood and of an extended timescale was hinted at only in private letters. It was evidently too early for a unified history of the Earth and life to be anything like successful.

Nevertheless, the preoccupation with the flood did provide a stimulating point of departure, from which research about fossils and rock strata could be carried forward. The works of Burnet and Woodward triggered a veritable flood of controversy, forming a debate that extended over the next few decades. However, they were also followed by a succession of theories of the Earth, which continued the discussion of Earth history in terms of natural process. These carefully reasoned proposals were

A fossil fish, figured in a book by Ray, who had little doubt that it represented the remains of a real fish and was not just some 'sport of nature'.

firmly based on evidence drawn from studies of fossils and sedimentary rocks, of rivers and oceans, of earthquakes and mountains and much else besides. Hence, through the agency of the flood, fossils did indeed become a key that opened a new interpretation of the history of life.

Within John Ray's lifetime, one more major puzzle for the naturalist was to raise its head. Building on his collaboration with Willughby, Ray spent many of his later years preparing handy synopses of the main groups of animals. In his analysis of the different groups, he was much influenced by, and in the case of the fishes actually helped by, an anatomist named Edward Tyson (1650–1708). Having studied at both Oxford and Cambridge, Tyson practised as a physician in London. There he taught human anatomy at Surgeon's Hall and was elected to the Royal Society. He made his reputation by applying his anatomical skills to the study of species other than man; his first major publication, in 1680, described the bodily organization of a porpoise in meticulous detail.

So when a young chimpanzee was shipped to London in 1698, arriving in a rather sickly condition, Tyson was the obvious choice to examine it while it still lived and to dissect it after the unfortunate creature had died. His results were duly reported to meetings of the Royal Society and then published as a monograph sponsored by the Society in 1699. Although monkeys were well-known in Europe at that time, having long been imported as pets, the apes were virtually unknown and Tyson's work was the first accurate description of such an animal. In the seventeenth century, the term 'ape' was used very loosely and the term chimpanzee had not yet come into use. Tyson called his animal, and his monograph,

Orang-Outang, sive Homo Sylvestris, a name adopted from Nicolaas Tulp, who had briefly described the external appearance of a similar animal in 1641. Tulp's animal came from Angola and must also have been a chimpanzee, though the name Orang-outang came from Malaysia and means 'man of the woods' or in Latin '*Homo sylvestris*'.

Tyson sought to interpret his results in terms of the ladder of nature. This concept, it will be recalled, originated with Aristotle, who acknowledged that some species seemed to represent intermediates between successive grades. Indeed, in the *History of Animals*, he described 'the ape' and noted how it shares the properties of man (a biped) and four-footed animals (the quadrupeds). The ladder of nature, Tyson informs us, consists of a graduated series running from minerals through plants and animals to man, and even on to higher spiritual beings such as angels. He thought that knowledge of this elegant gradation should give us a deep respect for the Creator.

However, this concept was rather ambiguous in its application to humans. On the one hand, it set out a clear hierarchy of creation with man placed well above the animals and well below the angels. This emphasized man's superiority over the 'brute beasts', which was the subject of a good deal of self-congratulation at that period. On the other hand, it suggested that there were no big gaps in the chain; each species was expected to provide an almost imperceptible link to the next so that the dividing line between animals and men should be indistinct. It was evidently the latter aspect that Tyson had in mind when he placed his chimpanzee in the chain as an 'intermediate link' between man and monkey.

The method that enabled him to place his animal correctly was to study the same parts of the body in his animal, in a monkey and in a man. We may observe 'nature's gradation in the formation of animal bodies', he explains, by comparing the bodies of separate species part by part in this way. His great skill in developing the comparative method was in itself an important technical advance. This method continues to be used to this day in the attempt to classify animals correctly. And it enabled Tyson to show that the chimpanzee resembles us more closely than any other creature that we know of. His main achievement, therefore, was not just the accurate description of the chimpanzee but also the point-by-point demonstration of the remarkable similarity between the chimpanzee's bodily structure and that of humans.

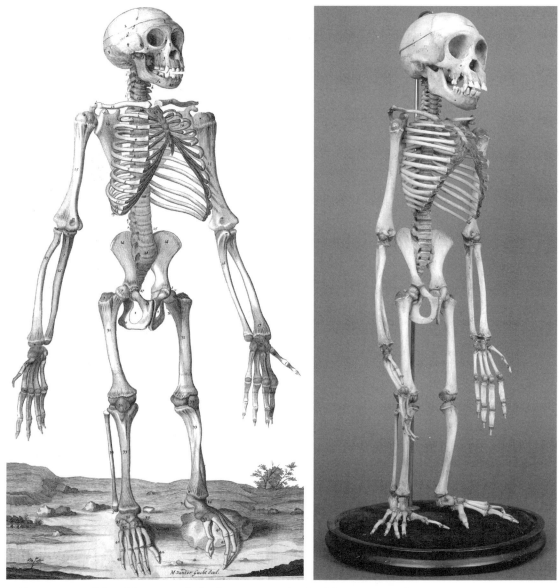

Every part of the chimpanzee's body was examined with great care by Tyson. He even noted correctly that the animal's skin is white beneath the dense covering of hair. The muscles, the genital organs, the brain and especially the skeleton were described and, where appropriate, precise dimensions of the parts were recorded. To make comparisons of the skeleton, Tyson obtained a monkey skeleton of similar size, as well as referring to the work of earlier naturalists. Then at the end of the monograph, he drew up a catalogue of the animal's features, noting 48 that resembled a human and 27 that resembled a monkey.

Tyson's comparative assessments are for the most part correct, bearing in mind that he examined an infant chimpanzee, which resembles a human more than an adult. It may also be the case that he unconsciously exaggerated the similarities to a human because the ladder of nature concept led him to expect a close resemblance. The animal's posture may provide an example of this. In discussing the chimpanzee's skeleton, he says 'we may safely conclude, that nature intended it a biped' just as humans are. Yet he does not discuss the significant differences between chimpanzees and humans in the structure of the pelvis and of the foot, both of which he described accurately. Tyson had seen the chimpanzee progress by walking on its knuckles, as these animals usually do, but he supposed that this was an unnatural posture due to its weakened state.

The intense interest in this work led to its being rushed into print. Although earlier thinkers had suggested that man stood close to animals on the ladder of nature, no one before had been able to point to a particular animal as evidence for this. Yet the result was not regarded as controversial because it appeared to confirm an established theory. In the long run, however, this study did sow the seeds of revolutionary change in our ideas about humanity's relation to the rest of the living world. The sense of separation between men and animals was already being eroded by a greater appreciation of domestic animals and the growing habit of keeping household pets, including monkeys. This trend was decisively reinforced by the revelation, through comparative anatomy, that human and animal bodies are so similar.

The close resemblance between ape and human was demonstrated beyond all possibility of doubt by Tyson's meticulous detail. This result was to exert a powerful influence on natural history over the following century. With Tyson's monograph on the chimpanzee, the puzzle of humanity's place in the scheme of things was implicitly added to the naturalists' agenda. This fact was not immediately recognized by contemporary naturalists because the new information was at first incorporated in the existing theory of the ladder of nature. Not until a century later would the concept of a linear and static ladder of nature be finally proved false. But the value of Tyson's work would remain even then.

This illustrates an important and fascinating aspect of science, namely that it is not necessary to hold a correct theory in order to do useful work. Useful questions may be tackled and accurate information accumulated by researchers whose underlying theory turns out to be quite wrong.

Tyson and his contemporaries helped prepare the way for evolution, not by somehow anticipating evolutionary ideas, but by tackling the right questions and developing sound methods within a set of ideas that were mostly incompatible with evolution. No one illustrates this situation better than the noted Swedish naturalist, Carl von Linne, generally known as Linnaeus.

CHAPTER THREE

MATTERS OF PLACE AND TIME

'Linnaeus and Cuvier have been my two gods', wrote Charles Darwin in later life. This praise for his predecessors was not because they anticipated some part of his evolutionary theory, for Linnaeus and Cuvier held views that were incompatible with evolution. What both men did was to contribute some key evidence and techniques in the study of the major puzzles confronting naturalists. These advances provided the foundations on which a scientific understanding of the origin of species could be built. By tracing out the line of thought that runs from Linnaeus to Cuvier, therefore, we take some important steps along the path that leads to the discovery of evolution.

When Carl Linnaeus (1707—78) turned his attention to the world of nature, he was faced with a growing number of species new to science. In the case of mammals, for example, he had to cope with twice as many species as Ray had listed half a century before. Linnaeus was not daunted by this multitude of species for he was a great organizer, who devoted his abundant energy to tackling the problems of classification. The solution he came up with was to devise a system for naming species and arranging them in groups, which was so straightforward that any naturalist could use it. More than any other method on offer, that of Linnaeus made it easy to identify and catalogue species of plants and animals. It was therefore received with enthusiasm and adopted by the great majority of eighteenth-century naturalists.

The method that Linnaeus developed was outlined in the *System of Nature*, published in 1735. This first edition contained only twelve pages of text but these were unusually large, and double-page spreads were used to present tabulated classifications of the mineral, vegetable and animal kingdoms. On the first page, Linnaeus presented a numbered list of 'observations', which set out his views on exactly what species are.

Each species, he explained, consists of similar individuals bound together by reproduction, in which eggs always produce offspring closely resembling the parents. From this, it follows that 'no new species are produced at the present time'. Linnaeus also noted that individuals tend to be multiplied in each generation. Working back from the present, this meant that the line of descent through reproduction must eventually come to a single set of parents for each species. The origin of this first pair of individuals he attributed to 'some omnipotent and omniscient being, namely God, whose work is called Creation'.

On this view, species are real units that have stayed constant in form from the moment of their creation, apart from trifling variations due to local conditions. Such a view was consistent with the Aristotelian tradition in natural history, which regarded species as well-defined entities, each separated from the next by certain essential features. The constancy of species was also supported by the observations of naturalists, which showed that species in any one locality are clearly separated from each other, and that they do not change significantly from one generation to the next.

This view was at variance with contemporary folklore, which accepted an extraordinary degree of fluidity in nature. There was much popular lore about one species of plant growing from the seeds of another, and about deformed births being the result of copulation between animals of different species. Only when naturalists had put these popular notions behind them could the study of species really make much headway.

So both the reality and constancy of species seemed to be confirmed by scientific observation, by traditional logic and by inferences about their creation. From this perspective, the naturalist's task was to discover the pattern of resemblances and differences between species and to represent this accurately in a system of classification. The result would be the 'natural system' of classification because it would group species in a way that reflected the order of nature. Achieving this was the goal towards which Linnaeus worked during his life. He recognized, as Ray had done, that a truly natural system would have to take into account many points of resemblance and difference between species and between larger groupings. No one character could be used to the exclusion of others in arranging the larger groups.

However, Linnaeus recognized the need for an 'artificial system' that classified plants on the basis of a single structural feature as a convenient method to be going on with. Drawing on the recent discovery of sexuality

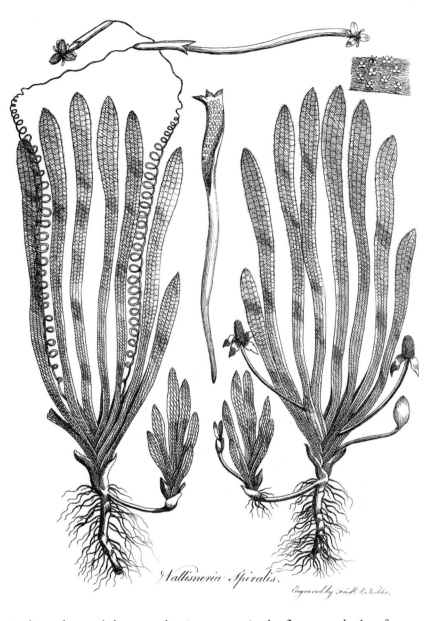

An illustration from The Botanic Garden *by Erasmus Darwin, who explained Linnaeus' method of classifying plants by the reproductive organs of the flower. The species shown is* Vallisneria spiralis, *which is now a popular aquarium plant.*

Vallisneria Spiralis.

Engraved by Fredk. P. Nodder.

in plants, he used the reproductive organs in the flower as the key feature on which his artificial system could be based. As set out in 1735, this method used differences in the number and position of the stamens and pistils. On this basis, Linnaeus was able to distinguish 24 major groups or classes within the plant kingdom.

In seeking to provide an integrating structure for his classification of plants, Linnaeus adopted a hierarchy of four levels: class, order, genus and species. The genus, which was already familiar to naturalists from

Aristotle's work, is an assemblage of similar species. Correspondingly, the order is an assemblage of similar genera and the class an assemblage of similar orders. Such a system is called an inclusive hierarchy because each level includes those below it: the plant kingdom includes the 24 classes, each class includes several orders, each order includes several genera and each genus several species.

The *System of Nature* introduced the same hierarchy of levels into zoology, where Linnaeus distinguished six classes within the animal kingdom. The differences between these six classes were sufficiently obvious that no artificial system was required for their identification. Classifying all living things into categories at these four levels gave the *System of Nature* a clarity and consistency that were absent from earlier systems of classification.

In later editions of the *System of Nature*, Linnaeus introduced the innovation for which he is best remembered. This is the practice of designating each species by two Latinized names, the first of which defines the genus, the second the particular species. Thus roses are placed in the genus *Rosa*; the dog rose is *Rosa canina*, the field rose is *Rosa arvensis*, and so on. This 'binomial nomenclature' had been used sporadically by earlier authors, such as Ray, but it was Linnaeus who standardized it so that every species of plant and animal was named in a consistent way. His standardized nomenclature was applied to plants first of all, and then extended to all known species of animals in 1758, with the 10th edition of the *System of Nature* (see colour plate 5).

Among the animal species named there was our own species, which Linnaeus designated as *Homo sapiens*. His descriptive notes covered human anatomy, morals, politics and racial differences. In his classification, there was a second species in the human genus, *Homo troglodytes*, but this proved to be a fanciful creature based on travellers' tales. Given the incomplete knowledge of the apes, there was naturally much uncertainty about the assignment of names. To the chimpanzee, described by Tulp and Tyson, he gave the name *Satyrus tulpii*.

However, Linnaeus was in no doubt that there were other species similar to humans in their anatomy. The genus *Homo* is included in the order Primates (Anthropomorpha in the first edition), along with apes, monkeys and lemurs, without any special comment. The primates, in turn, are included in the class Mammalia, as part of the animal kingdom. When he was criticized for classifying man with the brutes, Linnaeus defended

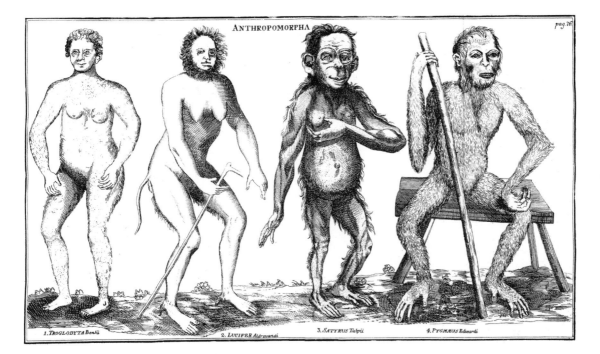

Four supposed 'manlike apes' classified by Linnaeus. The two on the left were the product of fanciful imagination. The third from the left, Satyrus tulpii, *is the chimpanzee and that on the right,* Pygmaeus edwardi, *may represent the orang-utan.*

himself by challenging naturalists to show him any structural feature by which the two could be separated more clearly. Moreover, he made no attempt to represent the pattern of genera or species as a linear ladder of nature with man at the top.

Taken together, the elements of Linnaeus' system helped to transform the practice of natural history. His firm concept of the species as a distinct natural entity reflected current scientific views and was an antidote to folklore. His binomial nomenclature provided a consistent way of naming everything from our own species to the smallest organism visible under the microscope. His higher categories of orders and classes also proved to be a satisfactory method for grouping species together on the basis of detailed comparative study. For the student of life's diversity, Linnaeus had developed a system of classification that was eminently workable and would last down the centuries.

At the same time, Linnaeus took a keen interest in what we would nowadays call the ecological relationships between species. He was well aware that animal and plant species are much more than just names in a catalogue. They also represent a living world, in which one species depends upon another for food, shelter or some other requirement. He had to consider this point in the light of growing evidence that each species is found only in certain parts of the world. This meant that both the geographical

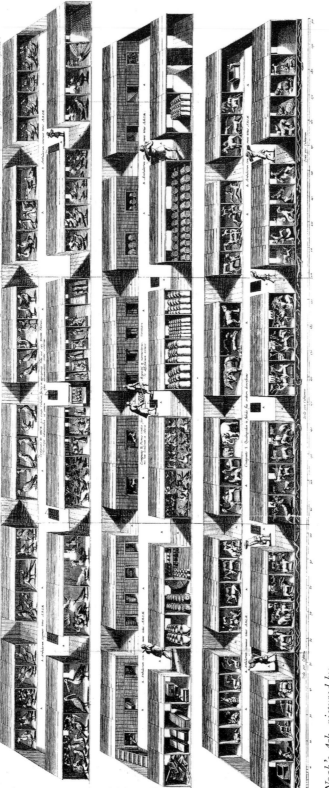

OPTICA PROJECTIOTHIUM ARCÆ NOÆÆQUÆ CONTINEBANTUR SPECIES CUNCTARUM ANIMANTIUM SPECIES PROPRIIS FIGURIS EXPRESSÆ NEC NON EARUNDEM STABULA ET MANSIONES EX UTRAQUE ARCÆ PARTE ORDINE DISPOSITÆ SPECTANTUR AMBULACRA IN ADMINISTRATIOREM

Noah's Ark, pictured by Kircher in 1675. Kircher envisaged the Ark as a large rectangular box crammed with compartments for the animals; he does not seem to have been troubled by considerations of seaworthiness.

distribution and the history of living things were drawn in to his discussion.

These discussions were contained in several essays, written over a number of years, some of which were defended as theses by his students. In the first of these essays, published in 1744, he tried to account for how the world has been stocked with animals and plants since the time of the flood. According to the story in Genesis, one pair of each kind of animal had been preserved in Noah's Ark during the flood, and these had returned to their normal lives as the floodwaters subsided. The trouble with this story from an eighteenth-century perspective was that the Ark was too small to have held all the species known to science.

The most elaborate calculation of how all animal species could have been accommodated in the Ark had been made in the previous century by the German scholar, Athanasius Kircher. It was his work that had helped to inspire the flood theories of Burnet and others in England. But his calculation was soon overtaken by the rapidly increasing number of species, and Ray had already dismissed Kircher's work on this topic as 'trifling and superficial'. To Linnaeus, it was simply incredible to think that every species of animal had been preserved in a wooden ship and had then repopulated the Earth a second time.

Instead, he telescoped the story of the flood into that of the creation, and proposed that all living things had their origin on a mountainous island surrounded by the primeval ocean. Located near the equator, this mountain was high enough to offer a range of climatic conditions according to altitude, just as mountains do today. The original pair of each species was then created in the appropriate climatic belt, from reindeer and Arctic lichens at the top to tropical palms and monkeys at the bottom. In this way, the mountain represented the entire Earth in miniature. Thereafter, each species increased in every generation and spread out over a larger area, as the primeval waters subsided. Just how organisms might have made the journey to their present locations was the subject of a full discussion in this essay.

Linnaeus went on to consider ecological relationships more closely in an essay entitled the *Economy of Nature* (1749). He developed the idea that each species of plant and animal forms part of a delicate balance of nature. In other words, each species is adapted to fulfil some role in the natural cycle of existence and so is needed to keep it in balance. Even such disagreeable creatures as maggots have a place in the economy of nature by consuming carcases and making their constituents available to other organisms.

In emphasizing that each species has its place in nature, Linnaeus also meant a particular geographical place. For each species is adapted to environmental conditions that are found only in certain parts of the Earth. Each kind of plant, for example, is adapted to a particular soil and a particular climate. 'Hence the same plants grow only where there are the same seasons of the year and the same soil', Linnaeus observes. So, in stressing how every species is adapted to its place in nature, this view emphasized the link between a given species and a particular geographical region. However, the closer the links between living things and their natural surroundings were seen to be, the less credible did the idea of migration from a single centre become.

This is what made the geographical distribution of species such a perplexing problem for eighteenth-century naturalists. It was generally assumed that each species had been created with a particular set of adaptations, which had remained constant over time. But if a species is adapted to conditions found only in a certain locality, then it could hardly have reached that locality by migrating across thousands of miles of hostile territory. And this difficulty applied equally to the story of Noah's Ark and to Linnaeus' primeval island. Naturalists therefore had their doubts about these stories as an explanation for the geographical distribution of animals, quite apart from the impossibility of fitting all the known species into the Ark.

An alternative approach was available in the work of George Louis Leclerc, Comte de Buffon (1707–88). As keeper of the King's Garden in Paris, Buffon set about producing a comprehensive survey of natural history in keeping with the encyclopedic spirit of the eighteenth century. The volumes of his *Natural History*, which began to appear in 1749, dealt with the Earth, the human species, domestic animals and a long list of wild animals. As well as the descriptive catalogue of species, the *Natural History* also contained bold theoretical essays and it was widely read by educated people across Europe.

Buffon differed sharply from Linnaeus in his approach. The root cause of this difference was that Buffon drew his inspiration not from Aristotle but from Descartes and Newton. He accepted their view of nature as a divinely created system of matter in motion, which is governed by divinely ordained laws. This meant that a truly scientific concept should be based

An illustration from Buffon's Natural History *showing the tarsier, recently described from the islands of South East Asia. Since Buffon declined to use the Linnaean system, it was left to later authors to allot this small mammal its place in the order Primates.*

TARSIER.

on the observation of physical objects that are linked to the rest of nature by physical causes and effects. So, instead of taking up what he saw as the vain search for abstract relationships, Buffon looked for 'physical truths' in his study of natural history. This was reflected in the volumes of his *Natural History*, which paid more attention to animals and plants in their natural surroundings and said little about their classification.

The geographical distribution of animals over the surface of the Earth was one of the issues that Buffon discussed from this point of view. What he did was to compare the species known from the Old World of Africa,

Asia and Europe with those described from the New World of North and South America. This comparison made it clear that many species of animals existing in the New World were different from those in the Old World, particularly in the tropics or 'torrid zones'.

It was obvious, for example, that the elephant, rhinoceros and hippopotamus lived in the continent of Africa but were not found in the Americas. It was also true that the big cats were different on the two continents, even though European explorers had confused the issue by giving familiar names to new-found animals. Attention to detail showed that the puma and jaguar were not misplaced lions or tigers but truly different species of big cats. In fact, asserted Buffon, the general rule was 'that not one of the animals of the torrid zone of one continent are natives of the torrid zone of the other'.

Buffon then tackled the question of why it was that the tropical mammals of the Old and New Worlds are exclusively confined to their own areas. He pointed out that tropical animals cannot maintain or reproduce their kind in colder areas. Hence they are confined either by the sea or by lands too cold to support them, and so cannot cross from one continent to another. This clearly implied that these sets of species could not have migrated from a single point of origin. They must have arisen in the areas where they are currently found.

Yet Buffon did allow migration some role in explaining the obvious similarity of certain animals living on either side of the Atlantic. In these cases, he supposed that the Old World forms may have migrated to the New World at an earlier period of the Earth's history, when the climate was warmer. They were then cut off by the cooling world climate and became changed into their new forms under the influence of local conditions. Such changes involved a degeneration from the original type, which had been created on the Old World pattern. Even humans had shown a corresponding degeneration on reaching America, in Buffon's view. But he was hard put to suggest just what factors caused this decline.

The general idea of the 'degeneration of animals' is discussed in a later volume of his *Natural History*, where he uses the big cats as an example. Though there are different species of big cats in the Old and New Worlds, all of them clearly belong together in the same 'family'. They closely resemble each other both externally and internally, and they show similar behaviour. The unity of the group is especially clear from the fact that the species on one continent differ more from each other

Plate 1A *The eastern grey kangaroo*, Macropus giganteus, *drawn in 1789 by George Raper, a midshipman with the First Fleet to Australia. He evidently found it difficult to interpret a mammal that naturally stands on its two hind limbs. Kangaroos illustrate the pattern of animal diversity, which is readily explained in terms of evolution.*

Plate 1B *The tree kangaroo*, Dendrolagus lumholtzi, *from North Queensland, Australia. This species is adapted to life in the trees by several features that distinguish it from ground-dwelling kangaroos, notably the relatively longer and stronger fore limbs. Such adaptations can be understood in terms of evolution.*

Plate 2A & B Two species of great ape illustrated in The Animal Kingdom *by Georges Cuvier. By the time this book was published in 1817, both the orang-utan (LEFT) and the chimpanzee (BELOW) were well known, but the gorilla had not yet been described. The discovery of apes so similar to humans raised questions about humanity's relationship to the rest of the animal kingdom.*

Plate 3A & B The crucial role of first-hand observation can be seen by comparing these two illustrations of a hyaena. ABOVE, the animal's scavenging habits are interpreted according to ancient folklore in this illustration prepared by a thirteenth-century illuminator for an English Bestiary. *LEFT, the animal's appearance is rendered accurately in a nineteenth century* Dictionary of Natural History, *and the animal is identified by its scientific name,* Hyaena vulgaris.

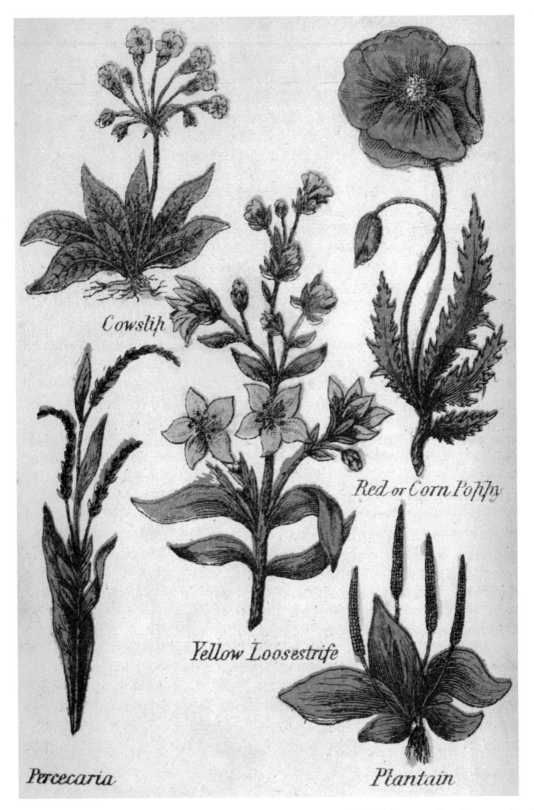

Cowslip

Red or Corn Poppy

Yellow Loosestrife

Percecaria

Plantain

Plate 4 Plate from The Complete Herbal by Nicholas Culpeper, published in 1649. The plants were listed more or less in alphabetical order and their medicinal properties were described.

The contrast between Old World (LEFT) and New World (RIGHT) species of big cats, discussed by Buffon, is shown in these illustrations from Thomas Bewick's History of Quadrupeds. *The African lion is quite distinct from the American lion or cougar (ABOVE), as is the leopard from the jaguar (BELOW).*

than from those on the other continent. From all this, we may justly conclude 'that these animals had one common origin and that, having formerly passed from one continent to the other, their present differences have proceeded only from the long influence of their new situation'.

It is difficult to be sure whether these ideas constitute an early theory of evolution. Buffon himself wavered in his opinion about whether truly new species are produced by these means, and in any case he usually spoke in terms of degeneration from a more perfect type. Nevertheless, he clearly thought in terms of a sequence of historical changes brought about by natural causes, after the manner of Descartes' philosophy. He also introduced the idea that a group of species, such as the big cats, is a real group because all its members are descended from a common ancestor. The line of descent constituted a sequence of natural causes that gave the group a physical reality.

On the same principle, Buffon accepted species as real entities because of the succession of individuals that reproduce among themselves and give rise to similar individuals. In fact, he insisted that the test of a true species is that the animals reproduce together. He knew, for instance, that bison are native to America and that the ox had been introduced there from the Old World. His own guess was that earlier oxen had migrated to America in warmer times and had turned into bisons under the influence of a colder climate. The differences between the two now seemed sufficient to place them in separate species but one could not be sure until the test of inter-breeding had been made. 'Nothing could be more easy,' he suggested, 'than for the inhabitants of Louisiana to try if the American bison would copulate with the European cow.'

Naturalists across Europe were not much impressed with speculations such as this on Buffon's part. As far as they were concerned, his ideas about the causes of distribution seemed to have no better foundation than those of Linnaeus. But even if his ideas were not well received, the accurate information in successive volumes of Buffon's *Natural History* certainly added to the knowledge of how species are distributed around the world. The geographical distribution of species was an aspect of diversity that was clearly as significant as the variety of species reflected in their classification.

As the eighteenth century went on, many of the younger naturalists were inspired to embark on long and dangerous voyages to collect specimens and study local geography. A number of them did not return. The work of those who made it safely home did indeed confirm that different parts of the world are inhabited by different species. More than that, it showed there are geographically distinct groups of species. In each group, the species are ecologically linked to each other as well as to their own geographic area. Evidently, each of these groups forms a coherent unit of plants and animals found in one geographical region only.

So, in the latter part of the eighteenth century, naturalists came to for-mulate the modern concept of regional 'floras' and 'faunas'. The flora and fauna of each region were seen as an ecological unit assembled from many species of plants and animals, respectively. The geographical region to which each unit is confined became known as a 'biological province'. And the recognition of these biological provinces implied that each assem-blage of species is in some way native to its region.

In turn, this meant that naturalists had to abandon the idea that all species had dispersed from one geographical source. It seemed much more reasonable to suppose that each species of animal and plant had been created in the area where it now lives. Consequently, the idea of a single centre of origin for all species was replaced by the idea of many 'centres of creation' at separate points around the world.

Ideas about the distribution of life based on Noah's Ark and the flood were thus put aside by the naturalists working on this topic. A related topic of much interest was the development of ideas about the history of life based on examining the rocks of the Earth. This challenge was taken up with enthusiasm by Buffon. He had begun the *Natural History* with a volume on the Earth, in which he sought to write a complete world history without even mentioning the flood. Later on he revised this part of his work and published it as a supplement to the *Natural History*. This volume, entitled *The Epochs of Nature*, appeared in 1778.

In the interval between these two editions of Buffon's work on the history of life, one particular piece of evidence began to exert a strong influence on contemporary discussion. This was the recovery of an increasing number of elephant-like fossils from the superficial gravels in Europe and Russia, along with similar fossils that were coming to light in North America. Eighteenth-century naturalists no longer regarded such large fossil bones as the remains of giant people who had lived in ancient times. They were able to recognize the resemblance of these bones, and the associated tusks and teeth, to those of elephants. The fact that living elephants are no longer found in either Europe or North America added to the interest of these fossils.

The finds that raised the most profound questions came from the Ohio Valley in the United States and belonged to what we would now call a mastodon. An initial collection of a few bones and teeth was shipped to Paris and received a meticulous study at the hands of Louis Daubenton (1716–1800), a young anatomist whose detailed investigations supported Buffon's project. Daubenton took the novel step of comparing the thigh bone (femur) of the animal from Ohio with the same bone in an elephant and in one of the Russian fossils, which we now know as a mammoth. This direct comparison showed that the three bones share the same distinctive anatomy despite differences in size and proportion. Accordingly, Daubenton concluded that the three animals represented by these bones all belonged to the same species.

This conclusion was reinforced by his study of the tusks. Although the tusks differed in size and shape, the fine anatomy of the ivory appeared to be similar in all three cases. However, Daubenton was baffled by the cheek teeth (molars) of the Ohio animal, which had knobs on the

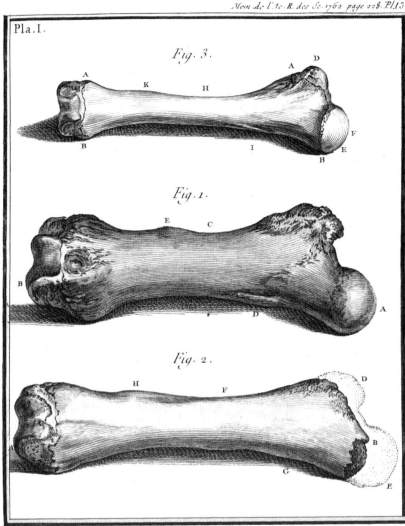

Mem. de l'Ic. R. des Sc. 1762 page 228. Pl. 13

Pla. I.

Fig. 3.

Fig. 1.

Fig. 2.

Ingram Sculp.

The femur of a mastodon from Ohio (1) compared to that from a Siberian mammoth (2) and an elephant (3). This remarkable comparison between fossil and living species was made by Daubenton in his 1762 paper on the Ohio fossils.

crown instead of the narrow cross-ridges found on typical elephant grinders (see illustration on page 71). He could only suggest that the teeth belonged to some other animal and had become mixed up with the bones and tusks of a giant elephant at the Ohio site.

This idea could not be sustained as more material was recovered from the Ohio Valley and, this time, was shipped to London. There the fresh finds were studied by English naturalists, who presented their results in papers read to the Royal Society of London in 1767 and 1768. They were agreed that the fossilized bones, tusks and molars from Ohio all belonged to one species of animal; these remains were found together too consistently to be an accidental association. The English naturalists were also agreed that this species was clearly different from the living elephant and was currently unkown, possibly even extinct. This was a possibility that Daubenton had found unthinkable.

The questions raised by this new material were submitted to Buffon in a letter from London. Could there really be a large animal with tusks like an elephant and bumpy grinders unlike those of any elephant? Could it be the case that such a species was now 'lost', that is, extinct? And how could one explain the fact that these fossils were found in a region with a cold climate, when today's elephants lived only in the tropics? These questions were tackled by Buffon in his *Epochs of Nature*. While he could not agree that the tusks and molars belonged to the same species, despite the evidence of association, he had no doubt that the bumpy molars belonged to a 'lost species'. Moving beyond these detailed questions, the mastodon and the mammoth also played a key role in the grand narrative that unfolded in the *Epochs of Nature*.

Buffon envisaged seven major epochs in the history of the Earth, which began as a globe of molten matter ejected from the sun, much as in Descartes' *Philosophy*. During the second epoch, the globe cooled and developed a solid outer crust, which is represented today by the lowest rocks. These lack fossils since the early Earth was unsuitable for life. Then in the third epoch the crust cooled to the point where water could condense, leading eventually to the formation of a worldwide ocean. Abundant marine life soon arose in the ocean, and these organisms left their remains in the lower sedimentary rocks.

The level of the ocean fell during the fourth and fifth epochs, resulting in the emergence of the continents and the appearance of life on land. At that stage, the climate was warm enough to permit tropical animals to exist all over the world, as shown by the fossil elephant bones of Europe and North America. These warmth-loving animals were gradually excluded from the northern regions as the planet cooled and today live only near the equator. In the sixth epoch, the separation of the continents took place with the appearance of the oceans, which prevented migration by large animals like elephants. Finally, the age of man constituted the seventh and present epoch.

How long had it taken for all this to occur? Buffon tried to answer this question by estimating how long it would take a molten globe the size of the Earth to cool to its present temperature. At the hamlet from which he took his name, Buffon had established an iron foundry. There he had his workmen prepare balls of mixed iron and non-metallic substances, in a graduated set of sizes. These were heated to near melting point and then allowed to cool to air temperature in a nearby cave. As expected, the larger the ball, the longer it took to cool. From these results, Buffon estimated that it would take a ball the size of the Earth about 75,000 years to cool. At least, that was his published estimate. In unpublished manuscripts, he toyed with the possibility that the Earth might be as much as 3 million years old. Either way, the acceptance of a much-expanded timescale was part and parcel of Buffon's approach.

The *Epochs of Nature* represents the culmination of the grand theories of the Earth that had prospered since Burnet's time. This succession of theories had built up the idea of an unfolding world history, which could be reconstructed by scientific inference. A new outlook was created, in which the successive rock strata came to be treated as historical documents of the Earth. Buffon's own use of fossil evidence to infer past changes, for example, shows a sound historical insight. But by the time this, his final work, appeared in print, the emphasis in science had shifted to the accumulation of precise facts and their subsequent explanation. So the amount of theorizing in the *Epochs of Nature* was felt to be unduly high in proportion to the evidence available.

As it happened, the right sort of evidence was being gathered by work that grew out of the earlier flood geology, which Buffon had scorned. The views of Woodward, in his *Natural History of the Earth*, had been well received on the continent of Europe early in the eighteenth century. Following Steno's initiative, he had emphasized that layered rocks with fossils must have been deposited sequentially at the time of the flood. This view also promoted practical research by encouraging the study of fossils and rock layers or strata. As time went on, less attention was paid to the flood and more to the actual sequence of rock strata. A generally agreed classification of rocks eventually came out of this detailed work.

The European champion rock classifier was Abraham Werner (1749–1817) at the school of mining in Freiberg. He drew together the work of earlier investigators to produce his *Short Classification and Description of the Different Rocks*, published in 1786. The idea that the Earth's crust contains an orderly succession of layers was placed beyond doubt by this work. Where these strata were undisturbed, the order in which they lay must be the order in which they were formed, with the

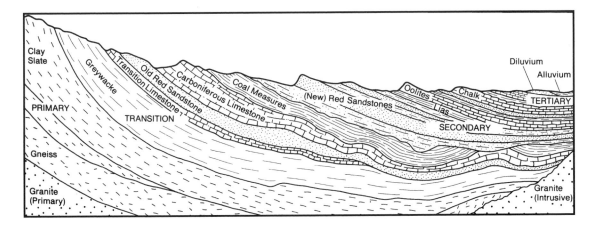

Diagrammatic section of the Earth's crust, showing Werner's rock classification with its major categories of Primary, Secondary and Tertiary. This section is based on one published in 1836 by William Buckland, in his Bridgewater Treatise on Geology and Mineralogy. *For part of Buckland's original figure see colour plate 7.*

oldest at the bottom. So Werner classified the rocks according to age, recognizing three basic categories, which were termed 'Primary, Secondary and Tertiary'.

The Primary rocks were usually crystalline and lacked strata; they contained no fossils but were often rich in ores. These rocks were attributed to an early period in Earth history. Overlaying them, and hence younger, were the 'Secondary' rocks, which were stratified and often rich in marine fossils. Then there were the 'Tertiary' rocks of still younger age, consisting of fossil-bearing clay and sand; these rocks were also stratified. Later, rocks were found that were intermediate between Primary and Secondary in age and other attributes; these made a fourth category, the 'Transition' rocks. In addition, the superficial sediments washed down from the mountains by rain and streams were distinguished as 'Alluvium'.

Werner interpreted this classification in terms of his own theory of the Earth. Like Buffon, he supposed that all the Earth had been covered by an ocean at an early stage of history. But he differed from Buffon in thinking that the primary rocks had been formed by chemical precipitation from that ocean rather than by cooling. A common element in the theories of both men was the idea of a retreating ocean. This idea was designed to deal with the crucial problem of how sedimentary rocks, formed by deposition from water, came to be found on dry land high above the sea level. The simplest explanation seemed to be that the sea was once deep enough to cover all the land and had since retreated.

The only available alternative was that the land had been lifted up by some internal force such as earthquakes or volcanoes. But Werner thought that volcanic activity was mostly quite recent and was due merely to the burning of coal beds and other subterranean materials. There was,

63

in his view, no force capable of elevating the Earth's surface. Because it was tied to a practical and efficient classification, Werner's theory proved popular with naturalists who studied rocks and minerals, and remained so until the early nineteenth century.

Evidence for the great extent of volcanic activity in the past was the stumbling block that eventually undermined Werner's theory. An early example of this kind of evidence was provided by studies on the Massif Central of France. This had been recognized as a region of extinct volcanic cones by a French naturalist returning from a visit to Vesuvius and nearby volcanic fields (see colour plate 6A). Subsequent analysis of this region in terms of its volcanic features was taken up in earnest by Nicholas Desmarest (1725–1815), beginning in the 1760s. He made detailed maps of the basalt masses in the Auvergne district, and showed how these resembled lava flows. Some of them he could trace back to the base of well preserved volcanic cones or even to the crater. This led him to argue that all sheets of basalt were of volcanic origin.

Desmarest went on to make inferences about the history of the region by noting details such as the relative position of the basalt layers and their degree of erosion. For instance, it was not unusual for him to find a relatively uneroded basalt layer at the bottom of a valley and another, highly eroded, basalt layer on the crest above the valley. This situation implied that there had been two volcanic eruptions widely separated in time, one occurring before and the other after the process of erosion that had carved out the valley. From all his evidence, Desmarest concluded that three episodes of volcanic activity must have occurred in the Massif Central, each separated from the next by a long period of erosion. He did not offer any suggestion as to just how much time was involved. Rather, his achievement was to demonstrate how a sequence of past events could be reconstructed by observing the arrangement of precise facts in the present.

Speculative writing on the history of the Earth continued to appear alongside the cautious fact-based work of men like Desmarest. One bold theoretical essay that had a lasting influence was published by James Hutton of Edinburgh. Having discussed his ideas in private for many years, Hutton (1726–97) read two papers at the newly formed Royal Society of Edinburgh in 1785. His work was duly published in volume 1 of the Society's *Transactions* in 1788, with the familiar title 'Theory of the Earth'.

In that essay, the central thread of his argument was that the Earth is a machine designed by God to sustain life. This purpose is partly served by processes of decay. The fertile soil, on which our food plants depend, is formed through the erosion of the rocks by wind and rain. However, this process of erosion will eventually wear down entire continents and render the Earth unfit for terrestrial life. Therefore there must be some way in which the world is rescued from decay, unless we are to conclude that the Earth was deliberately made imperfect.

Hutton maintained that evidence for a process of restoration is not hard to find. Most of the world's land is made of sedimentary rocks, which are composed of the sediment eroded from former continents. These marine sediments must have been raised to form new continents, and subterranean heat is the agency that has accomplished this. Hutton pointed out that the power of internal heat is clearly seen in action during the eruption of a volcano such as Mount Etna. Its effects are evident in the fracturing and contortion often seen in strata uplifted from the bed of the sea. And the large extent to which internal heat has acted in times past is clear from the evidence of ancient lavas and extinct volcanoes gathered by Desmarest and others.

So erosion was a force that could wear down the land, and internal heat was a force that could elevate it again. Decay and restoration form a natural cycle of events that keeps the world machine in balance. This makes it possible to explain the structure of the Earth in terms of causes that can be seen in action at the present time. Judging by the rate at which these forces act today, an immense period of time must have been required for just one of these cycles of decay and restoration. But then, granted time, Hutton saw no reason why this cycle of events should not be repeated indefinitely. A 'succession of worlds' is recorded in the rocks, and 'we find no vestige of a beginning, no prospect of an end'.

While waiting for his theory to appear in print, Hutton undertook some journeys around Scotland to gather factual support for his ideas. At Glen Tilt in the Grampian Mountains, he was excited to find granite in cracks (dykes) cutting almost vertically across the other strata. Again, on the Isle of Arran, he found a situation where granite had intruded into much older strata and forced them to arch upward. Such things would not be expected from precipitation in a primeval ocean. They demonstrated that granite is more reasonably thought of as 'unerupted lava'. When it cools slowly

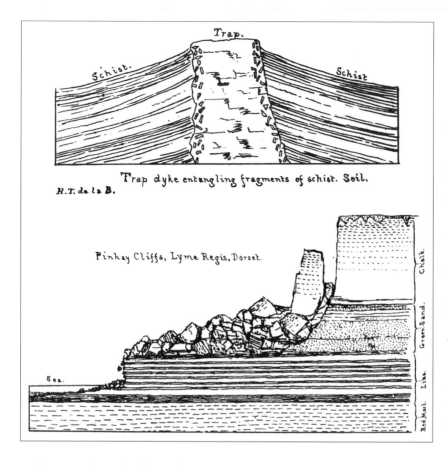

Trap.

Schist.   Schist.

Trap dyke entangling fragments of schist. Soil.

H.T. de la B.

Pinhay Cliffs, Lyme Regis, Dorset.

Chalk.

Green Sand.

Lias.

Red Marl.

Sea.

Two drawings showing the kind of evidence that supported a broadly Huttonian view of geological processes. ABOVE, a dyke formed by intrusion of molten rock; the material of the dyke (Trap) has detached portions of the rock through which it cuts. BELOW, cliff erosion by percolating freshwater, which undermines the porous Chalk and Greensand and carries them toward the sea over the more impervious clay (Lias).

underground, evidently molten rock can acquire the crystalline form that Werner had taken as evidence of precipitation from water.

Then in 1788 Hutton saw a remarkable situation in the strata of the cliffs at Siccar Point. Here the thick sedimentary strata are tilted at a steep angle, and are overlain by further sedimentary strata that are more or less horizontal. This represented a gap in the record of rock strata, now known as an unconformity, which was rendered conspicuous by the sharp difference in the angles of the strata. Hutton reasoned correctly that this situation implied a long sequence of changes. The tilted strata must first have been deposited horizontally by sedimentation and then tilted on end by subterranean upheaval. Next they must have been eroded and then sunk again beneath the sea, where the second lot of strata could accumulate over them. Finally, all these strata must have been elevated again to their present position.

This example provided dramatic evidence for the cycles of erosion and uplift that Hutton had in mind. His friend, John Playfair, who

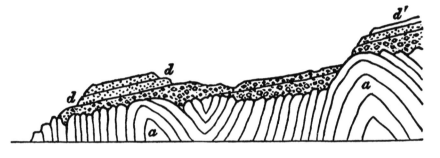

A view of Siccar Point, near St. Abb's Head in Scotland, showing the striking unconformity in the rock strata visited by Hutton. The diagrammatic section through the cliff (BELOW) shows the lower strata (a), which have been disturbed and tilted, and the upper strata (d), which have been deposited horizontally upon the lower. These illustrations are from a later textbook by Charles Lyell.

accompanied him on this outing, later recalled the deep impression that it had made. 'What clearer evidence could we have had of the different formation of these rocks', he wrote, 'and of the long interval which separated their formation, had we actually seen them emerging from the bosom of the deep?' Thinking over this sequence of events made it plain what a vast length of time must have been involved. As Playfair commented, 'The mind seemed to grow giddy by looking so far into the abyss of time.'

When Hutton's views were published, they added fresh controversy to the debates about the history of the Earth. He was taken to task by those who preferred Werner's views or who picked on his lack of practical experience in geology. Hutton responded to his critics by publishing an enlarged *Theory of the Earth*, as a two-volume book, in 1795. In the long run, of course, both Hutton and Werner can be seen to have contributed in important ways to this ongoing debate.

Hutton had provided good evidence for central heat as a geological force. As more evidence came in, the idea of dynamic interaction between the forces of erosion and uplift prevailed over ideas that relied on water alone. The way that Hutton was able to make sense of unconformities in the strata also gave strong support to a more dynamic view. In addition, this view implied that geological forces must have interacted over an enormous period of time.

Werner would not have quarrelled with this last point since he himself had remarked that 'our Earth is a child of time'. His main contribution was building up an accurate classification of the superimposed rock formations. This was based on detailed rock sequencing in many localities, a kind of fieldwork for which Hutton had little taste. Moreover, the investigations of people like Werner and Desmarest gave world history a direction, which was lacking from the repetitious cycles of Hutton's theory. Clear evidence for directional change, of the kind that Buffon and Werner had envisaged, was to come from the history of life.

To write a history of life, one must study fossils. The study of fossils soon received a masterly treatment at the hands of Baron Georges Cuvier at the museum of natural history in Paris. Cuvier (1769–1832) was born in a small French town, and he obtained a good education at an Academy in Stuttgart. After this, he passed the turbulent years of the French revolution as a tutor to a noble family in Normandy. This position gave him ample opportunity to pursue natural history, and he devoted much of his time to the animals of the seashore. These he studied by dissecting them with great care, a technique that he had learned at the academy. Then in 1795 the new government of France offered him a post in the capital, as a professor at the Natural History Museum.

Cuvier had developed his own approach to zoology through a careful study of Aristotle's books during the years in Normandy. Aristotle's concept of final causes meant recognizing that the parts of the body studied by the anatomist exist to serve particular functions in the life of an animal. But, Cuvier argued, a living body is not just a collection of organs for performing certain functions. The organs must be arranged, and their functions must interact, so as to produce a harmonious whole. The life of an animal depends not only on the performance of certain functions but also on their coordination. Therefore an anatomist must take into account the functions of parts of the body as he studies their structure.

Cuvier put this into practice by using two anatomical rules. The 'correlation of parts' was the first rule, which expressed the idea of functional coordination in the body. This rule stressed that an animal must have all parts of its body correlated with one another so as to fit it for a given way of life. The 'subordination of character' was the second rule, in which Cuvier argued that the most important parts of the body for classification are those that are least modified by adaptations to different ways of life.

Comparison was the method that Cuvier advocated for discerning the rules of correlation from among all the facts of animal structure. By comparing the same organ in different species, one can see how that organ is modified in the different contexts. Different species are like experiments performed by nature, which adds or subtracts parts, and the comparative anatomist can then see the results of these changes.

These basic principles were set out in the first volume of Cuvier's *Lectures in Comparative Anatomy*, which was published a mere five years after his arrival in Paris. The volumes of the *Lectures in Comparative Anatomy* did not follow the traditional pattern of describing the whole anatomy of one species after another. Instead, they dealt with one part of the body at a time, compared right across the animal kingdom. Volume one dealt with skeletons and muscles, volume two with nervous systems, three with digestive systems and teeth, and so on. This proved to be a most fruitful approach, the more so on account of the thoroughness with which the comparisons were made. Whereas previous works had compared several species, the *Lectures* compared hundreds.

Nor was it just a matter of quantity. The minuteness and accuracy of Cuvier's work set a new standard in comparative anatomy. The smaller invertebrate animals were described in as much detail as the larger, and more familiar, vertebrates. In volume four, for example, he endeavoured to trace the delicate blood vessels of molluscs and crustaceans as completely as those of vertebrates. The finer details were even clarified by injection of coloured fluids into the blood vessels. So Cuvier's anatomical work excelled in both quantity and quality.

From this wealth of information, Cuvier illustrated the correlation of parts by referring to the differences he had found between flesh-eating and plant-eating mammals. A flesh-eating animal must be able to see its prey, to catch it and to tear it apart. Hence it possesses legs adapted both for fast running and for grasping, as well as jaws with cutting teeth for tearing flesh. Cutting teeth are never found in the same species as a foot cased in horn that cannot be used for grasping. For hooves on the feet are correlated with a jaw bearing flat-crowned molar teeth for grinding food. All hooved mammals are plant-eaters.

In the *Lectures*, Cuvier went on to explain that this correlation of parts means that any one part of an animal is a reflection of the whole.

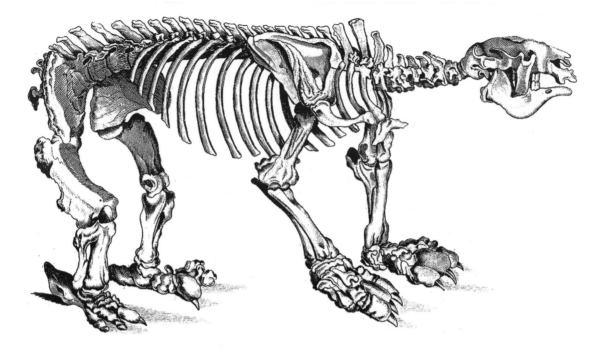

MEGATHERIUM (Megatherium)

An anatomist who finds certain functional modifications in one part of an animal may therefore predict the corresponding modifications that will be found in other parts. 'Consequently by looking at one single bone,' said Cuvier, 'the appearance of the whole skeleton can be deduced up to a certain point.' This claim was made good in the dramatic reconstruction of fossil vertebrates. Cuvier's interest in fossils began soon after he came to Paris. In 1796, he was asked to report on some giant fossil remains that had been sent to Spain from Paraguay in South America. Cuvier was given a set of engravings, from which he carefully described this creature and named it *Megatherium* (Latin for 'huge beast'). His comparative studies showed that, in spite of its great size, this animal belonged to the same family as the living sloths of South America. But no gigantic living sloths were known to exist. Therefore it seemed almost certain that the *Megatherium* must be an extinct species.

This encounter with *Megatherium* fired Cuvier's interest and led him into a brilliant series of studies on fossil animals. He saw that the reconstruction of fossil vertebrates could be accomplished with the aid of his anatomical rules. The *Megatherium* had been almost complete but most fossil vertebrates are found only as scattered and disarticulated bones.

The giant fossil skeleton from South America, which Cuvier named Megatherium, *illustrated in his book on* The Animal Kingdom. *It was Cuvier's study of* Megatherium *that set off his interest in fossils and the possibility of extinction.*

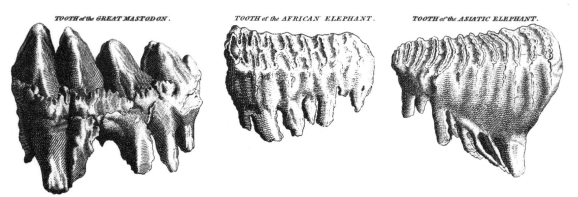

TOOTH of the GREAT MASTODON.   TOOTH of the AFRICAN ELEPHANT.   TOOTH of the ASIATIC ELEPHANT.

The molars of a mastodon from Ohio (LEFT) compared to those of an Asian elephant (RIGHT) and an African elephant (CENTRE) in Blumenbach's Natural History. *Cuvier made a fresh study of these teeth in his effort to settle the question of extinction.*

Often the bones of more than one species are represented in any given deposit. The risk of reconstructing a bogus animal from such a mixture could be avoided if one realized that each bone formed part of a functionally integrated whole. By correlating each bone with the rest, in relation to the whole skeleton, Cuvier could be sure of matching up those bones that truly belonged together.

He determined to settle the question of extinction by sorting out the living and fossil elephants. Since elephants are so large, one could be reasonably sure that no more living species remained to be discovered in some remote spot. Hence if the fossil species were found to be different from the living elephants, they must be extinct. From the material in the Paris collections, Cuvier was able to show first of all that the African and Asian elephants are separate species. He found differences in the vertebrae, the number of ribs, the length of the tusks and especially the molar teeth. The differences in the pattern of cross-ridges on the molar teeth and in the other anatomical features were too great and too constant to be variation within one species.

In the same way, he was able to show that the fossil 'mammoth' was distinct from either of the living species of elephant. Johann Blumenbach in Germany had independently come to much the same conclusion about this fossil form. The first complete skeleton of a mammoth from Siberia became available about this time, and Blumenbach considered that this was identical with the fossils from Europe. He even gave the mammoth a Latin name in the Linnaean classification, *Elephas primigenius*. Cuvier agreed, and he accepted Blumenbach's name for this species. It seemed beyond reasonable doubt that this mammoth must be an extinct species.

In addition to this, Cuvier's full report of 1806 took the American fossils into account. Some of the fossils found in the New World were definitely mammoths, he decided. But the elephant-like fossils that had been sent

71

from the Ohio Valley were something else. These finds had perplexed earlier naturalists because the limb bones and tusks were similar to those of elephants but the molar teeth were quite different. Cuvier used his methods to show that the bones, tusks and teeth all belonged to one species, which he termed a 'mastodon' on account of the remarkable teeth. Clearly, this was another extinct form.

These results established the fact that some species had become extinct. This was a most important demonstration because, as Cuvier himself pointed out, it was a truth about the past based on 'facts of observation' in the present. It was possible to read back from the present into the past by means of these meticulous comparisons between the anatomical details of fossils and those of living species. By the use of these comparative methods, Cuvier had been able to recognize the mammoth and mastodon as distinct species of elephants, which had evidently vanished from the surface of the Earth.

Hitherto, the idea of extinction had been resisted for a mixture of philosophical and religious reasons. For one thing, it would make gaps in the great chain of being, which would reduce this elegant chain of creation to a shapeless chaos. Although Linnaeus had ignored the chain of being for practical classification, it continued to exert a strong hold on eighteenth-century thought. Another problem was how an entire species could be driven to extinction in a world ordered by divine providence. Surely this could not happen in a balanced economy of nature?

But extinction held no philosophical or religious terrors for Cuvier. He plunged into a detailed study of other fossil bones in the superficial deposits around Paris. By careful comparison with the living forms, he discovered that he was dealing with extinct species of deer, elephant, hippopotamus, rhinoceros and others. Many of these mammals were larger than their living counterparts. It was not just the mammoth and mastodon, but a whole fauna that had been swept off the face of the Earth.

In addition, Cuvier searched for fossils in the gypsum quarries of Montmartre near Paris. These were Tertiary deposits, considerably deeper, and therefore older, than the gravels. Here he discovered some mammalian fossils that appeared to have no living counterpart at all. Rather, they showed a combination of characters, which are found separately in the living pig, rhinoceros and tapir. There were several similar species of different sizes, for which he made a new genus, appropriately named *Palaeotherium* (Latin for 'ancient beast'). It began to look

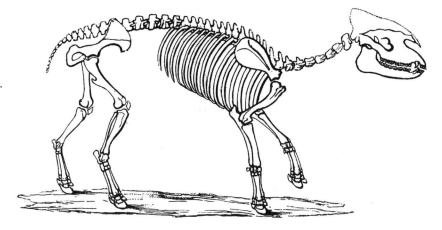

Cuvier's reconstruction of the skeleton of the early Tertiary mammal, Palaeotherium, *illustrated in Buckland's Bridgewater Treatise on* Geology and Mineralogy.

as if there was an extinct fauna from the Tertiary gypsum of Montmartre that was as diverse as the extinct fauna from the gravel. By 1812, Cuvier had accumulated enough material to fill a magnificent four-volume synthesis, *Researches on the Fossil Bones of Quadrupeds.*

Here, indeed, were exciting results with far-reaching significance. Cuvier had worked out sound methods for the reconstruction of fossil animals. The application of these methods showed that the present age had been preceded by earlier ages when animals that are now utterly extinct had flourished. The older the deposits in which these extinct animals were found, the more they differed from any alive today. As to how long ago these ages were, Cuvier never pretended to be precise, but he did speak of 'thousands of centuries' for the fossils from Montmarte.

Meanwhile, the style of thinking about species that Buffon had promoted was continued by others. In England, the subject was taken up by Erasmus Darwin (1731–1802), a prominent scholar and grandfather of Charles Darwin. His busy practice as a physician did not stop him exploring many other interests, ranging from steam engines to the conduct of female education. Botany was among these interests and in 1789 he published a poem, *The Botanic Garden*, which popularized Linnaeus' sexual system of plant classification for an English audience (see illustration on page 50). Between 1794 and 1796, he produced a two-volume treatise with the title, *Zoonomia; or the Laws of Organic Life*, in which the idea of evolutionary change in living things was explicitly discussed.

He gave examples to show that there is variation within a species, and he argued that change is driven by the forces of lust, hunger and security.

Darwinia lejostyla

A shrub of the genus Darwinia. *This genus was named in honour of Erasmus Darwin by Edward Rudge in the* Transactions of the Linnean Society *in 1815. The genus* Darwinia *is found only in Australia.*

For example, hunger 'has diversified the forms of all species of animals', as shown in the varied beaks of birds, which are adapted to take different kinds of food. These forces therefore interact with variation to bring about the improvement of species through time. All such change proceeds 'by its own inherent activity' based on the general animation of life. In these terms, Darwin sketched out the idea of an evolutionary origin for species, but the idea was not pursued in detail.

Just how far he was willing to go with this idea was made clear in his poem, *The Temple of Nature*, published posthumously in 1803. Here he suggested that all living things might have arisen from 'forms minute' on the ocean floor. This would have come about through the action of natural laws 'Impressed on Nature by the Great First Cause'. As a result, these organisms would have acquired new powers and more complex structures as successive generations passed. 'Whence countless groups of vegetation spring, And breathing realms of fin, and feet, and wing.'

Thus Erasmus Darwin put forward the general idea of evolution, and evolution on the grand scale at that, for all to see.

In France, Buffon's style of thinking was continued by Chevalier de la Marck (1744–1829), who was one of Cuvier's colleagues at Paris. Lamarck, as he was commonly known, had begun his career in botany, and achieved a notable success with *The French Flora* in 1778, but this book does not contain a word about evolution. Lamarck came to his theory of evolution only after changing his study from botany to zoology. And then he put forward his ideas, not in support of Cuvier's conclusions on extinction, but in an attempt to refute them.

CHAPTER FOUR

A NATURAL HISTORY OF CREATION

Lamarck's stature as a botanist is illustrated by an incident that occurred in 1789. The Committee of Finances of the new revolutionary government of France threatened to abolish his position as an economy measure. He responded by publishing a pamphlet entitled 'Considerations in favour of Chevalier de la Marck', which set out his accomplishments as a botanist and defended the usefulness of his position. His position was retained!

Not long after this, the government appointed Lamarck to a new position at the museum of natural history in Paris. He became the professor of 'insects, worms and microscopic animals'. Unabashed by being required to classify a different set of organisms, Lamarck soon became expert at his new post. He introduced a number of distinctions that have stood the test of time, such as that between vertebrate and invertebrate animals. He was also the first to distinguish insects, crustaceans and arachnids as separate classes. His first book on zoology, the *System of Invertebrate Animals*, appeared in 1801. This was well received, and one reviewer praised his system as 'certainly the most perfect that has yet appeared'.

It is this book that contains the first account of his newly formed evolutionary views, set out in a theoretical appendix to the main work. Lamarck was never content to remain a mere classifier of plants and animals. He liked to grapple with the widest of theoretical problems and he proudly labelled himself a 'naturalist-philosopher'. In this role, he was among the first to draw a clear distinction between inanimate objects and living things. He did this as early as 1778, in the introduction to his *French Flora*. Hitherto, three kingdoms had been recognized, those of mineral, vegetable and animal, as in Linnaeus' *System of Nature* of 1735. Natural history had dealt with the classification of the various natural objects in all three kingdoms.

But Lamarck and others saw the need for a separate science of living things. This would be concerned with those features that plants and animals have in common and that distinguish them from minerals. The new science should have its own concepts and methods, free from the view that living things could be analysed just like inanimate objects. And it should try to develop broad theories of life, in contrast to Cuvier's narrow and analytical approach to the study of animals. In 1802, Lamarck coined the term 'biology' for this new science of life.

The science of biology, as Lamarck saw it, included the idea of transformation as a central element. And he believed that the groups within the animal and vegetable kingdoms could be arranged on a linear series of increasing complexity. He saw this ladder of nature or chain of being as a true historical series produced by evolution. Nature had formed all organisms through evolution, beginning with the most simple and continuing progressively to the most complex. There was a steady movement of organisms up the ladder of nature, over time, due to an inherent tendency to increase in complexity. This tendency he attributed to 'the power of life'.

The upward movement, Lamarck supposed, must leave gaps at the bottom of the series. These gaps are made good by the spontaneous generation of 'animalcules of the simplest organisation' from inanimate matter. As soon as these are formed, they join the main series and begin to move upwards. Man is placed atop the series, representing the limit of evolution. Yet living matter does not accumulate at the top for human beings are subject to death and decay. In this way they return to the mineral state and become available again for fresh occasions of spontaneous generation. Hence Lamarck's theory is one of pre-programmed evolution, rather like movement up an escalator.

Lamarck acknowledged that neither the species nor genera of animals or plants could be arranged in a straight line. Only the main groups of animals and plants, such as large families or classes, could be arranged in a single series of increasing complexity. He thought that this departure from the expected linear series was due to a second evolutionary factor, that of adaptations brought about by local circumstances. Lamarck's views on how this second factor works were explained in his famous *Zoological Philosophy* of 1809.

Here, he states that when an animal's environment changes, it produces corresponding changes in the animal's organization in the course of time. He explains that alterations in the environment lead to alterations in the animal's needs. These in turn lead to alterations in the animal's 'activity', 'efforts' or 'habits'. Altered habits, such as the more frequent use of a given part, produce bodily changes that are inherited by the animal's offspring provided the changes are common to both parents. Such inherited effects of activity or habit were later dubbed the 'inheritance of acquired characters'.

This is not an idea that Lamarck originated, nor even one for which he claimed special credit, as it was widely accepted in his day. What was new with Lamarck was the incorporation of this idea into a theory of progressive evolution. According to his theory, the needs created by a changing environment are experienced by a kind of inner feeling, the '*sentiment interieur*'. This is a power of unconscious reaction to external stimuli found in animals that have central nervous systems. The unconscious reaction is able to direct 'vital fluids', which promote changes in certain parts of the body to meet the need. These changes are then inherited by the next generation.

In other words, Lamarck considered that adaptation is the result of an interaction between an organism and its environment. The environment influences an animal's habits, and the altered habits cause heritable changes in the body. This approach stood the argument from design upon its head, as Lamarck was well aware. According to the design view, nature or its Author has endowed each species with a constant structure that suits it to certain habits. But in Lamarck's view, a change in habits triggered by the environment produces a changed structure to suit the new circumstances.

A full and final account of his ideas on evolution was given in 1815 in the introduction to his *Natural History of Invertebrate Animals*. Nevertheless, his concept of evolution made little headway with his contemporaries. This was partly because Lamarck's views were often vague and inconsistent, which made it easy for his critics to dismiss them as unscientific, and it was partly because other issues came in to complicate the whole question of biological transformation.

Lamarck's belief in evolution developed in the context of the lively debate over extinction at the turn of the century. It was clear from his own work that, while some species of fossil shells were the same as living ones, there were many that were no longer to be found alive in areas so far explored. The question of the day was what had happened to these latter species. As Cuvier pointed out, there seemed to be only three alternatives: either they had evolved into the living species or they had become extinct or they had migrated to regions not yet adequately explored.

Both Cuvier and Lamarck agreed that migration might account for the differences between the fossil and living forms to some extent. But they disagreed completely about the other two alternatives. Concerning most of the species that could not be identified with living forms, Cuvier held that they had become extinct and Lamarck held that they had evolved into living forms. The two men debated the issue in terms of the relative merits of evolution versus extinction as explanations for the disappearance of fossil species.

Cuvier thought that extinctions were caused by huge natural disasters sweeping suddenly over part of the globe. Since each species was well adapted to a particular set of circumstances, most would be unable to cope with such a great change in their environment. But Lamarck refused to accept that there had been any such sudden interruptions to the ordinary course of events. He held that the processes of nature had acted gradually and continuously over a vast period of time. As evidence against any sudden revolutions, he pointed to the fact that many fossil shells belonged to species that are still living. Of those that did not, many belonged to genera still represented by similar living species. Therefore such changes as had occurred could not have been so very radical.

So Lamarck doubted whether any species had become extinct except for some of the larger mammals that had perished through the agency of man. He preferred to think that species were not stable units and that the fossil species had evolved into the living ones. But given such a choice between evolution and extinction, the evidence certainly seemed to support Cuvier rather than Lamarck, for as the fossil evidence continued to come in, it showed that Lamarck had been wrong in denying extinction, not that he had been right in advocating evolution.

Furthermore, Cuvier's work showed that there was no chain of being or ladder of nature for organisms to move up in the course of time as Lamarck's theory proposed. Using his anatomical rules, Cuvier set out to produce an improved classification of the animal kingdom. He saw that the most useful features for producing a broad classification are those that are most constant across large groups of animals. In practice, he found the nervous system to be the least variable part of the body and therefore the most useful in classifying the basic groups of animals. Cuvier's revised classification of major animal groups was first published in 1812. This was followed by the full results of his efforts in a four-volume work, *The Animal Kingdom*, published in 1817.

In the introduction to this book, Cuvier explained that his methods show that 'there exist four principal forms, four general plans, if it may be thus expressed, on which all animals appear to have been modelled'. The four forms are the vertebrates (Vertebrata), the molluscs (Mollusca), the arthropods (Articulata) and radially shaped animals (Radiata). In each of these four forms, the various parts of the body are coordinated in quite a different way. Each one represents quite a different design for living. Cuvier called each of these forms a 'branch' of the animal kingdom.

Recognition of these four distinct groups meant that the chain of being was finally broken. Never again would it be possible to arrange animals in a linear series. The cephalopods (squid, cuttlefish), for example, are such complex molluscs that it would be hard to place any other animal between them and the fish on a linear scale of complexity. Yet molluscs and vertebrates have next to nothing in common in their organization. It is only within each major group that a gradation can be found and, even then, these forms cannot be classified in a straight line from the simplest to the most complex. Having first published his conclusions in 1812, Cuvier was content to dismiss a linear classification in the *Animal Kingdom* with a few caustic references to 'the pretended chain of beings'.

However, this was not the end of the matter. One of the few pieces of positive evidence that Lamarck brought forward in support of his ideas came from the discovery of rudimentary teeth in the foetus of the minke whale (rorqual), which is a toothless baleen whale as an adult. These foetal teeth are small and never cut the gum, but are re-absorbed without ever coming into use (see colour plate 8A). Lamarck cited this case of a rudimentary structure as evidence for evolution. His colleague at the Paris museum, Geoffrey Saint-Hilaire (1772−1844), saw it primarily as evidence of 'unity of plan', although he was not hostile to evolutionary ideas.

The idea that all animals were constructed on a common plan had been put forward in the latter part of the eighteenth century. It was closely linked to the idea of a chain of being, and had formed an important part of the romantic nature-philosophy of the period. Under this influence, Saint-Hilaire tried to found a comparative anatomy based solely on the principle of unity of plan. In deliberate contrast to Cuvier's approach,

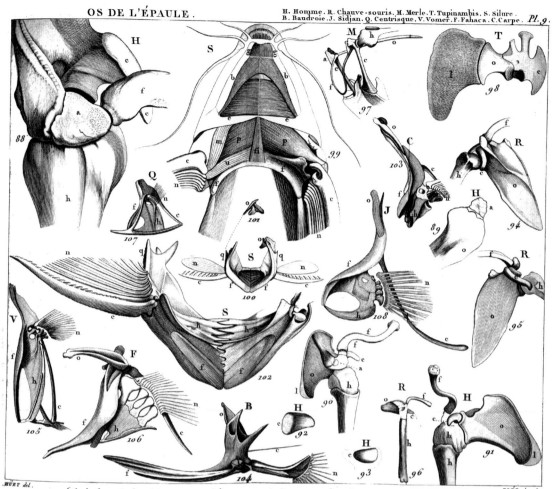

OS DE L'ÉPAULE .

H. Homme . R. Chauve-souris . M. Merle . T. Tupinambis . S. Silure .
B. Baudroie . J. Sidjan . Q. Centrisque . V. Vomer . F. Fahaca . C. Carpe . Pl. 9 .

(Clavicules . f. *furculaire* . c. *coracoïde* . a. *acromion*.) o. *omoplate* . l. *omolite* . h. *humerus* . n. *nageoire* .

A comparison of the bones of the shoulder in different vertebrates, from Geoffroy Saint-Hilaire's Anatomical Philosophy. *In each figure, the species is indicated by a capital letter (e.g.* C = carp, H = human), *and the bones are identified by lowercase letters (e.g.* c = coracoid, h = humerus).

this was to be a science of pure structure, without reference to functional considerations. He applied his ideas by comparing the structure of corresponding parts of the skeleton in different vertebrates. His results were published in a series of memoirs, and enlarged in his *Anatomical Philosophy* of 1818.

In one memoir, he compared the shoulder girdle and forelimb of land-dwelling vertebrates with the pectoral girdle and fin of fish, noting the corresponding bones in each. In another memoir, he compared the bones of the skull in different species. He was able to show that similarities that were not apparent in the adult forms could be made out by studying the embryos. In adults, the skull of fish is made up of a greater number of parts than that of crocodiles or mammals, but this difference disappears when the embryonic skulls are examined. When the bones are compared

81

in terms of the number of embryonic centres of ossification, the correspondence between the different skulls is a close one. Then it can be seen that 'the skull of all vertebrate animals consists of about the same number of parts, and that these parts always keep the same arrangement and the same relationship'.

These elegant results supported the idea that there is a definite unity of plan among the vertebrates. Not content with this, Saint-Hilaire embraced the idea that there is one common plan for all animals. By some ingenious arguments, he maintained that each part of an insect's skeleton is represented by a corresponding part in vertebrates. He also tried to argue that there is a similar correspondence between the organization of a vertebrate and that of a mollusc. This was based on the work of two young naturalists, who likened a cephalopod to a vertebrate bent double at the navel, in a paper presented to the Academy of Sciences in 1830.

All this was too much for Cuvier. Later that year, he presented a paper to the Academy, in which he launched an attack on Saint-Hilaire and the recently deceased Lamarck. With his extensive first-hand knowledge of both cephalopods and vertebrates, Cuvier had no difficulty in demolishing the hasty comparisons drawn by the naturalists (see colour plate 8B). He pointed out that, even where the two groups have similar organs that fulfil similar functions, these organs are differently arranged with respect to one another and are often constructed in a different way.

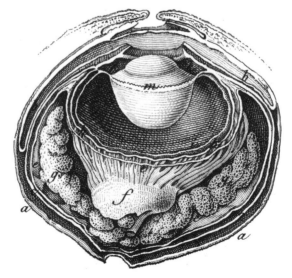

Cuvier's dissection showing that the eye of an octopus differs from that of a vertebrate in important features, despite their superficial similarity. For instance, the octopus eye has a bipartite lens (m), which has no muscles to change its shape, and an optic ganglion (f) enclosed within the outer coats of the eye.

As for Saint-Hilaire's unity of plan for the whole animal kingdom, it simply does not exist.

In this Cuvier was surely right. Nevertheless, Saint-Hilaire's work marks a step forward in the study of animal diversity. His results showed that there is a unity of structure among the vertebrates, which goes beyond the design requirements stressed by Cuvier. There was evidently more to the problem of animal diversity than just a multitude of designs for particular ways of life. So later workers in France and England began to adopt the best points from the work of both Cuvier and Saint-Hilaire. They recognized that there are distinct major groups, which differ in their organization, but they also accepted that one could trace a common plan within each major group. In taking this approach, they were stimulated to search for general theories of animal organization.

This approach also received support from the study of embryonic development. In 1828, Karl Ernst von Baer published a major work on animal development, in which he summarized his results in the form of four 'laws'. He had observed the actual course of development in many species under the microscope. After fertilization of the egg, he saw a series of complex events, from which the shapes and structures of the future adult gradually emerge. The egg divides to form a mass of cells, in which folds are formed, and these slide over each other, roll up and buckle to form the basic organs of the body. From this basic outline of the body, details are progressively refined and take on the form of the structures that will characterize the adult.

In short, von Baer's four laws describe development as a process of progressive specialization away from a general ground plan, which is laid down in the early embryo. Hence an embryo can be recognized early on as that of a vertebrate and only later on as that of a bird. The pattern of early development is not the same for all groups of animals. For example, early development is similar among all molluscs but this pattern is quite different from that found in vertebrates, which are also similar among themselves. So von Baer's meticulous studies showed that the four major groups defined by Cuvier are as distinct in the early embryo as in the adult. But his results also showed that within a major group there is a common ground plan from an early stage of development.

These debates over classification, anatomy and development were all part of the growth of biology as a distinct science. At the same time, questions about the history of life were also strongly influenced by some trends

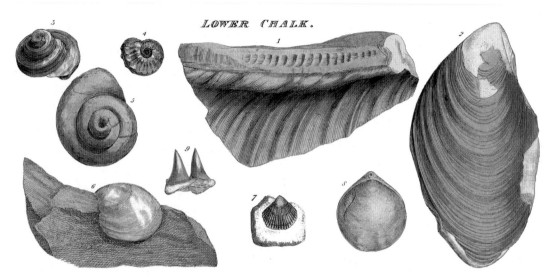

Cretaceous fossil shells from Smith's Strata Identified by Organised Fossils of 1816. The fossils consist of a variety of molluscs, except for the shark's teeth (9).

in geology, which came together and reinforced one another in the first quarter of the nineteenth century. As a result, geology matured rapidly in this period to become an exciting and highly professional field of study.

Among these trends was a growing acceptance that the forces acting on the Earth's crust interact dynamically in the way that Hutton had proposed. Even those who scorned the whole debate between Werner and Hutton tended to find evidence that supported a broadly Huttonian view of geological processes. Then there was the expanding study of fossils (palaeontology), which had been put on a sound footing by Cuvier. Increasingly, this was revealing that the Earth had been populated in past ages by species quite different from those now living.

Another development was the use of fossils to help in establishing the correct sequence of rocks and their correlation over a wide area. This technique, which built on the methods of Werner, received its definitive form at the hands of William 'strata' Smith in England and Cuvier's colleague Alexandre Brongniart in France. As these workers examined their local sequence of rock formations in detail, they found that each formation contained distinctive species of fossils. In fact, a close study of the fossils often helped them to discriminate between formations that were otherwise hard to tell apart.

Smith quickly saw the potential of fossils as a help in identifying specific formations and tracing them across country. The thoroughness of his work established the fact that each sedimentary rock formation contains a distinctive assemblage of fossil species wherever it is found within a given region. He used his skill in cartography to produce a series of geological maps, in which each formation was indicated by a different colour. And this work culminated in a huge coloured map of England and Wales, which he published in 1815.

The growth of this crucial technique, called 'stratigraphy', enabled geologists to establish local rock sequences and to correlate them across the countries of Europe and eventually worldwide. The advent of stratigraphy did more than anything else to turn geology into an historical science and to demonstrate the vast extent of past time. Together with the other advances in geology, this technique made it possible to begin an accurate history of life, tied to a history of the Earth as a whole.

Cuvier had begun to develop such a history in his work with Brongniart, and his ideas were set out in a 'Preliminary Discourse' to his *Researches on the Fossil Bones of Quadrupeds*. This was an important and readable outline of Cuvier's ideas, which was promptly translated into English and issued as a separate book in 1813. The English translation was given the old-fashioned title, *Essay on the Theory of the Earth*, a phrase that Cuvier had studiously avoided.

For evidence of what had caused the extinction of successive species of vertebrates, Cuvier turned to the rock formations of the Paris Basin. Some of these contained fossil bones of vertebrates along with fossil shells belonging to genera typical of fresh water. In between these formations there were others that contained fossil shells of marine organisms. It seemed that the formations of the Paris Basin represented alternate freshwater and marine deposits, and this implied some kind of alternation in the relative levels of land and sea. Cuvier concluded that recurrent flooding by the sea must have been the agency that caused the extinctions of the terrestrial vertebrates.

The transition from one condition to the other seemed to have been quite sudden, and this led Cuvier to envisage long periods of calm punctuated at intervals by sudden changes. These catastrophic changes, or 'revolutions' as he termed them, must have been due to some orderly natural cause. Whatever that cause was, he differed from Hutton in thinking that it could hardly have been any of the ordinary geological processes that we see at work today. Cuvier knew that strongly deformed strata were to be found in the foothills of the Alps, and he took this as evidence for violent dislocations of the Earth's crust. He supposed that these events had caused the fluctuations in sea level, with the resulting extinction of many species.

A lecture on geology at Oxford in 1823. The room is adorned with the characteristic objects that gave this new science its power: geological maps and sections hang behind the speaker (William Buckland), and a variety of vertebrate and invertebrate fossils lie before him.

Cuvier was willing to allow that the last of these violent revolutions could correspond with the flood recorded in the book of Genesis. In England, this view was taken up by William Buckland (1784–1856), who used it to defend geology against the charge that it undermined the credibility of the Bible. Buckland was the first person to be appointed as a reader in geology at the University of Oxford. He was a gifted lecturer, and many of his students acquired a lasting interest in geology. As evidence for the flood he pointed to the occurrence of superficial layers of gravel and mud in situations where they could not possibly have been laid down by rivers. He termed these deposits 'Diluvium' to distinguish them from ordinary Alluvium found in association with rivers.

The most dramatic examples of Diluvium were found in caves. Beginning with Kirkdale Cavern in north Yorkshire, Buckland visited many caves and found deposits rich in broken bones belonging to extinct species of mammals. The damaged state of the bones he attributed to their having been gnawed by an extinct species of hyaena, the bones of which were included in the deposits. He checked this point by observing the habits of a living hyaena in a zoo, and even kept one at home, much to the discomfort of visitors to his household. The possibility that early humans had lived in the caves and gnawed the bones did not occur to Buckland. He identified the layer of mud overlying the bones as sediment left by the flood.

Rather than a general submergence of the land below the sea, Buckland preferred to think of the flood in terms of some huge tidal wave. This explanation had already been invoked to account for deposits of gravel, and furrows in the underlying rock, around Edinburgh. This view was also supported by other evidence, such as large boulders transported far from their site of origin and U-shaped valleys unrelated in form to the rivers flowing in them. Later, all these phenomena were shown to be due to the action of glaciers. In the 1820s, however, the tidal wave theory of the most recent geological revolution seemed reasonable enough even to those who did not follow Buckland in his reconciliations with the Bible.

Cuvier paid close attention to the location of the vertebrate fossils within the gravels and rock strata. In his *Essay on the Theory of the Earth*, he noted that fossil species are related to particular strata in a way that is 'very distinct and satisfactory'. It was clearly ascertained, he wrote, that fossils of reptiles 'are found considerably earlier, or in more ancient strata', than those of mammals. Large fossil reptiles were coming to light at the top of the Secondary formations, in and below the Chalk, but there were no mammals there. This exciting harvest of discoveries included the first partial skeleton of what would later be called a dinosaur, which was named *Megalosaurus* by Buckland in 1824. In the same year, the marine reptile *Plesiosaurus* was fully described in a paper by Buckland's colleague, William Conybeare.

Fossil bones of mammals were only found well above the Chalk, in the Tertiary rocks and the Alluvium. Among these fossils, Cuvier pointed out, there is 'a very remarkable succession in the appearance of the different species'. Species that differ greatly from any living form, such as those in the genus *Palaeotherium*, are found in the more ancient Tertiary rocks, while extinct species that resemble living forms, such as those in the genera *Mastodon* and *Magatherium*, are confined to the Alluvium. This trend in the fossil record was rounded off by the fact that 'no human remains have been hitherto discovered among the extraneous fossils'.

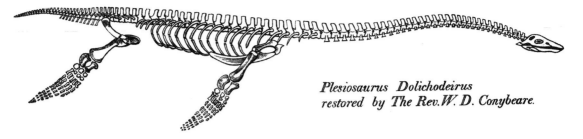

Plesiosaurus Dolichodeirus
restored by The Rev. W. D. Conybeare.

With this last point, Cuvier went a little too far. His assertion that 'there are no human bones in a fossil state' was not strictly correct even then. During the 1820s, flint tools and some human bones were being found in association with the bones of extinct animals, especially in caves. But both Cuvier and Buckland brushed these reports aside as due to inaccurate work or to some accident of preservation. They were not willing to recognize that the human race might stretch back in time to the era of the extinct species. However, this did not affect the valid point that the fossil record of vertebrates shows a broadly progressive trend. It began to be widely agreed that fossil species come to resemble living forms more closely as one passes from older to younger formations.

The fossil marine reptile, Plesiosaurus, *illustrated in Buckland's* Bridgewater Treatise on Geology and Mineralogy. *The restoration was done by Buckland's colleague Conybeare, who had found this fossil form in the Secondary (Jurassic) strata of England.*

Such a view received additional support from studies on fossil plants by Adolphe Brongniart (son of Cuvier's collaborator). In his *Forerunner of a History of Fossil Plants*, first published in 1828, he showed that the early Secondary formations of Europe were dominated by large ferns and other simple plants (cryptograms). Since similar forms are now characteristic of tropical areas, he concluded that the climate of that period was hotter than at present. In the later Secondary formations, plants such as pines and firs (gymnosperms) came to predominate over the cryptograms. Then in the Tertiary rocks the more complex flowering plants (angiosperms) appeared and dominated the scene. Each of these main floral periods seemed to be sharply separated from the next. The overall picture painted by the younger Brongniart was one of a progressive increase in the complexity and diversity of plants through time.

This approach culminated in the work of Elie de Beaumont, who used the new stratigraphical methods to date the episodes of mountain building in Europe. What came out of this was that mountain building had occurred at many different times, which often coincided with abrupt changes in fossil species. He suggested that the mountains could have been elevated suddenly by the release of stresses in the Earth's crust. These events would have produced the kind of tidal waves that Buckland had in

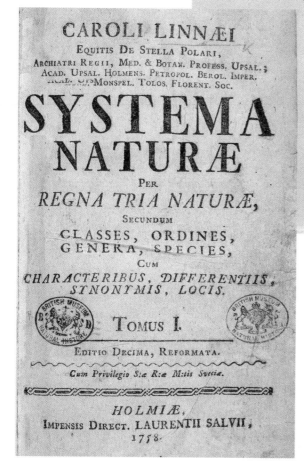

CAROLI LINNÆI

EQUITIS DE STELLA POLARI,
ARCHIATRI REGII, MED. & BOTAN. PROFESS. UPSAL.;
ACAD. UPSAL. HOLMENS. PETROPOL. BEROL. IMPER.
ACAD. U MONSPEL. TOLOS. FLORENT. SOC.

SYSTEMA NATURÆ

PER
REGNA TRIA NATURÆ,

SECUNDUM

CLASSES, ORDINES,
GENERA, SPECIES,

CUM

*CHARACTERIBUS, DIFFERENTIIS,
SYNONYMIS, LOCIS.*

TOMUS I.

EDITIO DECIMA, REFORMATA.

Cum Privilegio S:æ R:æ M:tis Sveciæ.

HOLMIÆ,
IMPENSIS DIRECT. LAURENTII SALVII,
1758.

Plate 5A & B The classification of plants and
animals by Linnaeus. ABOVE, details of the lady's
slipper orchid, from a painting by the famous botanical
artist, Franz Bauer. Using his binomial
nomenclature, Linnaeus gave this orchid the scientific
name Cypripedium calceolus. LEFT, title page of
the 10th edition of Linnaeus' System of Nature,
published in 1758, in which his binomial nomenclature
was applied to animals as well as plants.

Plate 6A & B Two volcanic localities that helped to reveal the vast extent of geological time. ABOVE, a valley among the volcanoes of central France, from Scrope's 1827 Memoir *on the region. The village stands on an ancient lava flow, and a cone and crater can be seen in the background. BELOW, the Bove Valley near Mt Etna in Sicily, which made a deep impression on Charles Lyell; a plate from his* Principles of Geology.

Plate 7 Part of a fold-out plate showing a diagrammatic section of the Earth's crust, from Buckland's Bridgewater treatise on Geology and Mineralogy. Rock formations typical of the Secondary era are shown in colour (BELOW) and their characteristic fossils are illustrated by the small drawings of animals and plants (ABOVE). For a simplified version of Buckland's diagram, see illustration on page 63.

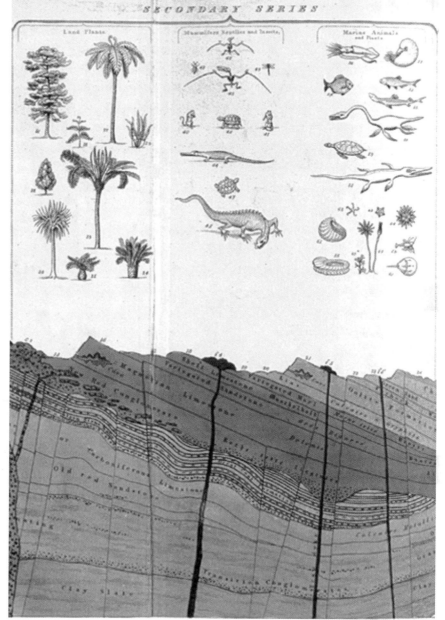

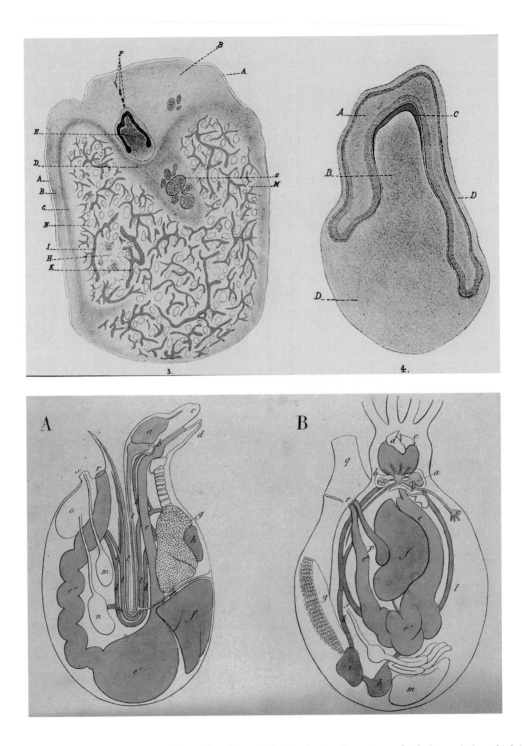

Plate 8A & B ABOVE, drawings of a section through the developing lower jaw of a baleen whale, which is toothless as an adult. Most of the section (LEFT) is filled with bone but an embryonic tooth is present in the upper part. This tooth is shown enlarged in a separate drawing (RIGHT). BELOW, diagrams prepared by Cuvier to show the difference in organization between a vertebrate (A) and a cephalopod (B). Equivalent organs in the two types are labelled with the same letter of the alphabet (eg. a = brain; f = liver).

A fine specimen of Plesiosaurus hawkinsii, *acquired by the Natural History Museum, London, in 1834. This species was described and named by the anatomist, Richard Owen.*

mind, and these would certainly have had catastrophic effects on the animals and plants living at each period.

So Cuvier's idea of successive revolutions had grown to provide a synthesis that made good sense of the available evidence. This synthesis embodied both the idea of a strongly directional history of life and the idea that this history had been punctuated by sudden revolutions. Although there was no linear chain of being, the evidence of fossils suggested that there had been a progressive development of the vertebrates from extinct forms in the older formations to the living species of today. This development seemed to have taken place in distinct episodes. The transition from one stage to the next was sudden and was due to geological revolutions that had resulted naturally from the physical constitution of the globe.

But then, in 1830, a young man by the name of Charles Lyell (1797–1875) published the first volume of his *Principles of Geology.* This was the most influential book on geology to appear during the first half of the nineteenth century. In it, Lyell not only set out to summarize the geological knowledge of his day, but he also launched an attack on the prevailing synthesis in geology. He was so successful in these objectives that the *Principles of Geology* became the focal point of a long debate about geological processes and the history of life.

Lyell's interest in geology had been fired by attending Buckland's lively lectures as an undergraduate at Oxford and by reading his father's copy of

89

Bakewell's *Introduction to Geology*. This was one of the first English pop-ular texts to come out strongly in favour of Huttonian views. It did not even mention the book of Genesis. As Lyell began his career in geology, he turned increasingly against his old teacher's views. He was much influenced in this new direction through reading a book on volcanoes by George Scrope, which appeared in 1825.

Scrope (1797–1876) was fascinated by volcanoes, and had set off at the age of twenty to study them in and around Italy. There he had the good fortune to witness the eruption of Vesuvius in 1822. He also went to central France to study the extinct volcanoes previously described by Desmarest. His work reinforced that of Desmarest, and convinced Scrope himself that the valleys there could not have been carved out by a giant tidal wave. Fortunately the area had not been affected by glaciers and so the evidence that the valleys had been excavated by gradual erosion remained clear.

Basalt layers, representing ancient lava flows, are found at different levels from near valley floors up to the caps of small hills in areas dissected by erosion. Scrope realized that the level of the basalt must indicate the level of the valley floor at the time of eruption. Where the basalt occurred high up along one of the present valleys, the river had evidently cut a new valley alongside the hardened lava flow. Hence the river now ran along a bed that was sometimes hundreds of feet below the earlier one preserved beneath the basalt. He pointed out that, even if one ascribed the first valley to the action of a flood, the second one could not possibly be accounted for in this way. The presence of loose volcanic scoria and tuff showed that no violent event had disturbed the landscape since the last eruption.

The landscape that Scrope described was open to no other interpreta-tion than that of the interaction of volcanic action and gradual erosion over an enormous number of years. Lyell saw that the next question was whether or not the extinction of the fauna had been as gradual as the excavation of the valleys. He set off for France to see the Massif Central for himself, accompanied by a friend. Lyell was confident that his own investigation would uphold the views of Scrope and he was not disappointed.

They found a rich fossil fauna in the river gravels preserved under the basalt layers. These fossils had just been described by local naturalists, who showed that the mammalian fossils were all extinct species of genera still in existence. They included elephants, rhinoceros,

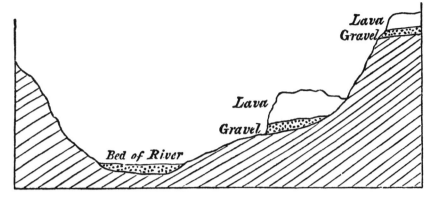

Diagrammatic section through a valley in Auvergne, in the Massif Central of France, showing ancient lava flows (basalt) at different levels above the present valley floor. At each eruption, the lava would have flowed down the existing valley floor, and this is confirmed by the layer of river gravel trapped under each layer of basalt. See also colour plate 6A.

hippopotamus and other species belonging to Cuvier's fauna of the Alluvium. Yet these were found in river gravels high above the present valley floor and so were very old when judged by the amount of erosion that had occurred since their deposition. This convinced Lyell that there had indeed been enough time for this fauna to have become extinct as slowly and gradually as the valleys had been excavated.

Lyell continued his journey southwards to Sicily, where he looked at the effects of volcanic activity round Mount Etna. The remarkable Val de Bove, which he visited at Buckland's suggestion, offered the opportunity to view what is in effect a vertical section through the mountain (see colour plate 6B). This revealed that the core of Etna is composed of many sloping layers of lava and tuff, showing that it has been built up by successive eruptions. There was nothing to suggest that the older lava flows were any greater than the recent ones. Hence the mountain must have grown slowly over a vast period of time.

When Lyell examined the fossil content of the strata upon which Etna stands, he found that the great majority of fossil shells belong to species still living in the Mediterranean. By geological standards, therefore, Mount Etna must be of recent origin, comparable in age to the most superficial deposits. Clearly, then, events that seem sudden on a geological scale can become resolved into a long sequence of unexceptional events when examined against a human timescale.

Lyell returned home determined to write a book that would show how all geological revolutions could be ironed out in the same way. The biological and geological changes of the past could have occurred by processes as gradual as those we see today. On top of that, Lyell resolved to set out the right 'principle of reasoning' that would make geology a science rather than just natural history or speculation. He considered that geology would only be a real science if it dealt with true causes, meaning those that could be observed in operation at the present time. This was made clear in his book's long subtitle, 'An attempt to explain the former changes in the Earth's surface by reference to causes now in operation'.

At his father's insistence, Charles Lyell had been trained in the law so that he might have some means of supporting himself other than inherited wealth. So when he came to write the *Principles*, it was natural that it should exhibit the skilled advocacy of the bar rather than the impartial judgement of the bench. He began the first volume with a history, in which geology was said to have been engaged in a 'violent struggle' against 'ancient doctrines'. Development of the subject was 'retarded' by the 'many prejudices with which earlier geologists had to contend'. This highly coloured account was meant to ensure the final separation of geology from the book of Genesis. It was also a device to make it appear that his opponents' views 'retarded' the progress of science while his own approach was 'the road that leads to truth'.

But within this context, Lyell made some valid criticisms of the whole idea of geological revolutions. The main source of error, he argued, is our failure to grasp just how vast is the duration of past time. He pointed out that even quite ordinary geological events will appear sudden and catastrophic if viewed on too short a timescale. In much the same way, the affairs of a nation would seem romantic and super-human if they were supposed to have occurred in one hundred years instead of two thousand.

Much of volume one was then spent on positive evidence that present geological processes could have produced all the observed effects in the past. He gave a skilled and comprehensive summary of the known effects of a whole range of geological agencies. These included erosion and silting, volcanic eruption, elevation and subsidence due to earthquakes, and so on. Here Lyell made good use of the first-hand experience gained on his travels through France and Italy, as well as showing his wide knowledge of the geological literature. He assembled all these examples in such a way as to emphasize their tendency to balance out over long periods of time.

As part of his overall strategy, Lyell also questioned the directional view of the history of life. Previously, he had agreed with his fellow geologists that there is a progressive trend in the fossil record of vertebrates. But this is simply an illusion, he asserted in the *Principles of Geology*, because the fossil record is too poor a sample of past life for us to draw any such conclusions. The apparent progression from reptiles to mammals is due to accidents of preservation. In any case, he argued, some mammals have been found in a Secondary formation, the Stonesfield Slate near Oxford, and this 'is as fatal to the theory of successive development, as if several hundreds had been discovered'. Thus he argued that there was no progressive trend in the history of life.

However, with his concern for processes, Lyell paid special attention to the way in which successive species become extinct in the course of time. In order to deal with this question, he plunged into a discussion of species and their ecological link to the environment. This discussion filled the second volume of the *Principles of Geology*. Here Lyell confronted the possibility of evolution as proposed in Lamarck's *Zoological Philosophy*, a book he thought was more like a novel than a work of science. He refuted Lamarck's ideas at some length but, in doing so, ironically made the concept of evolution more widely available to an English-speaking audience.

Lyell brought together a good deal of evidence to show that species are real units, which are well adapted to a given place in nature, much as Linnaeus had proposed a century before. He also maintained that species are stable units with strictly limited variability. Consequently, their adaptation will be upset as geological processes gradually change the environment, with the result that species will become extinct from time to time. So the piecemeal extinction of species should be 'part of the constant and regular course of nature'.

The extinct forms must then be replaced by new species, which are adapted to the changed environment. This introduction of new species was the subject of some shuffling of the feet on Lyell's part. He concluded that it must be as piecemeal as extinction even though he was quite unable to say how it comes about. Lyell seems to have been content to dodge around this issue since his main concern was to show that fossil species make reliable geological markers. But it was an obvious weakness in his position, and a few years later it caught the attention of a young man named Charles Darwin.

In the third volume of the *Principles*, Lyell went on to apply his ideas to the reconstruction of Tertiary formations. He provided good evidence that the Tertiary rocks must represent a vast period of time, during which

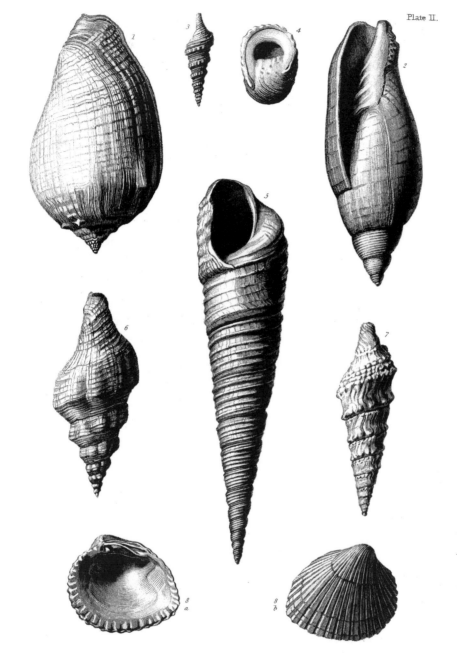

Plate II.

Some of the fossil mollusc shells characteristic of Lyell's 'Miocene' period, based on the work of the French geologist Deshayes. From volume three of the Principles of Geology.

fossil species have gradually come to resemble those of the present day. For one thing, as more formations were examined, many were found to be of different ages from those already known, judging by fossil content and other identifying features. Hence the passage from one formation to another was bound to become steadily less abrupt as more formations were discovered. Cuvier's sudden transitions were therefore an illusion due to incomplete sampling, and this point was reinforced by studies of the fossil molluscs.

Within the Tertiary period, Lyell distinguished four major formations, which he named Eocene, Miocene, older and newer Pliocene. Each of these contained species of fossil mollusc that were found only in that formation, but each also contained some species that were still living today. The proportion of still-living species decreases as one passes from the newer Pliocene to the Eocene formations. It was evident, therefore, that there was a steady turnover of species. Those characteristic of one formation gradually disappeared in later formations, just as a particular generation is found to a diminishing extent in successive census returns.

At the end of volume three, Lyell felt able to conclude that his geological theories were valid at least for the whole of Tertiary time. And he suggested briefly how these theories could be extended to the earlier and less well studied epochs. Taken together, the three volumes of the *Principles of Geology*, first published between 1830 and 1833, represent an impressive achievement. Lyell's single-handed efforts had not only summarized effectively the present state of geology but also proposed a new synthesis comparable with the established one that he opposed.

In retrospect, Lyell's synthesis in the *Principles* can be seen to contain three distinct elements, even though he himself made no distinction between them. The first of these elements is the method of reasoning back from the present into the past, assuming the constancy of natural laws. The second element is the argument that geological processes are the result of observable causes, which have never acted with greater intensity than at present. The third element is Lyell's claim that it was impossible for geology to discern any directional trends in the history of the Earth or of life because too much of the evidence had been lost. Logically, these are three separate ideas, but Lyell lumped them all together under the principle of 'uniformity'.

Not all of these three elements fared equally well in the debates following the first appearance of the *Principles*. As to the first of them, the method of interpreting the past in terms of present geological processes, this was already accepted by all of Lyell's colleagues. His opponents did not deny the soundness of this method, still less did they question the continuity of natural laws in geology. As Conybeare remarked in a review, no real scientist doubted that the past geological causes were similar at least in kind to those acting today. In spite of Lyell's polemical flourishes, there was really no disagreement over this method, which is essential for research in geology or any other historical science.

However, the idea that past changes have always been similar in intensity to those happening in the present is not part of an essential method. It is a particular claim about the past, which needs to be tested by evidence gathered in the field, as Lyell's critics recognized. They felt that some geological phenomena could only be explained by processes of a different order of magnitude to those acting at the present time. But on this point, Lyell carried the day and convinced his fellow geologists that sudden revolutions were not required by the evidence.

Part of the problem here was the difficulty of grasping the full implications of the vast geological timescale. By the 1830s, all participants in the debate agreed that long periods of time were involved. Buckland spoke of 'millions of millions of years'. And Conybeare suggested a figure of a 'few quadrillions of years' for the fossil-bearing rock formations, which is much too much even by modern standards. And yet, he could still argue that major geological events must have happened more suddenly and violently than in recent times. He failed to grasp the ironing-out effect that the timescale he accepted must have on apparently sudden events recorded in the rocks. Lyell clearly understood this point and illustrated it in successive editions of the *Principles*. Consequently, the cumulative weight of his argument and examples was most persuasive.

Another problem was that some things did not readily fit Lyell's view, such as those phenomena soon to be explained in terms of glaciers. At first, Lyell rejected the glacial theory when it was put forward by the Swiss naturalist, Louis Agassiz. The main reason for this may be that Agassiz (1807–73) linked his theory to Elie de Beaumont's theory of the sudden elevation of the Alps. Buckland, however, had no such reservations and was soon convinced that glaciation explained the effects that he had previously attributed to a great tidal wave.

When Agassiz visited Britain in 1840, he and Buckland went out glacier hunting together and found plenty of tell-tale traces of former glaciation in the Scottish highlands. A little later, Buckland was able to convince Lyell that there had been glaciers in Britain. 'On my showing him a beautiful cluster of moraines within two miles of his father's house,'

A glacier and its associated rocky debris (moraines). This figure is taken from a later edition of Lyell's Principles, *where he recognized the major effects of past glaciation.*

Buckland recalled, 'he instantly accepted it, as solving a host of difficulties that have all his life embarrassed him.'

The discovery of more extensive glaciation in the past illustrates how opposing views on the tempo of geological change became less divergent with time. The glacial theory was soundly based on reasoning from present processes back to past effects, and the advance and retreat of the ice was gradual, exactly as Lyell advocated. Yet the vision of a sheet of ice that overwhelmed Europe, carving out valleys and eliminating a whole fauna, was decidedly more catastrophic than anything in the first edition of the *Principles*. On the whole, then, Lyell was successful with the second element of his synthesis, but it was a mellowed and more flexible concept of gradual change that was generally accepted.

The third element in Lyell's synthesis, his denial that the fossil record gave evidence for a progressive history of life, was decisively rejected. Here he lost the support of friends like Scrope, who thought he had gone too far. Scrope recognized that the continuity of natural laws is not violated in the least by supposing that the Earth has 'passed through several progressive stages of existence'. Lyell could not see this, and so he felt bound to deny that there was any historical direction, even in the fossil record. This claim truly astonished other geologists since the weight of evidence was so much against it. Conybeare was moved to remark, in a letter to Lyell, that 'one of us must wear coloured glasses'.

By the time that Conybeare wrote this letter in 1841, the evidence had indeed made Lyell's position untenable. This evidence came out of the effort to define a sequence of rock formations that would have general validity. To this end, it became customary to define groups of formations that could be recognized from one area to another. These groups were called systems, and proved to be a useful unit for classifying rocks at a level intermediate between local formations and the broad categories of Primary, Secondary and Tertiary. At the top of the Secondary series, a 'Cretaceous' group was defined, which could be recognized readily by the distinctive chalk deposits. The bottom of the Secondary series was marked by a 'Carboniferous' group, which had been named by Conybeare and Phillips in 1822. This group included the distinctive and economically important coal seams.

Below the Carboniferous formations came the Transition rocks, which linked the Secondary series to the Primary rocks. The difficult Transition rocks were a challenge to the new stratigraphical methods, but they also held out the prospect of significant rewards. One of these was the possibility that they would contain the earliest fossils and so enable the history of life to be carried back to the beginning. Among those who determined to sort out the Transition rocks were two British geologists, Roderick Murchison and Adam Sedgwick. During their summer field trips in the 1830s, they worked in a region of suitable rocks on the border between England and Wales.

Murchison found an area where the known sequence of Carboniferous formations could be traced downwards into the Transition series. These strata contained an abundant and distinctive fossil fauna made up of invertebrates such as trilobites. There were no vertebrates except in the youngest strata. Murchison formally named this group the 'Silurian System', after a tribe from the Welsh borderland, in 1835. Meanwhile Sedgwick, in North Wales, had found a sequence of rocks that evidently came below the Silurian. In a joint paper with Murchison, published in 1836,

The armoured, jawless fish, Cephalaspis, characteristic of the Devonian period but making their first appearance in the fossil record late in the Silurian. The relatively short duration of the cephalaspid's existence is shown in Agassiz' diagram on page 100.

he named this the 'Cambrian System', after Cambria, the Latin name for Wales.

Then, in 1839, Sedgwick and Murchison designated additional formations between the Carboniferous and Silurian as the 'Devonian System', after limestone exposures in the county of Devon. These strata, which contain fossils of strange armoured fish as well as invertebrates, were the subject of a great controversy before their place was settled. But the validity of the Devonian System was confirmed when Murchison travelled across Europe and into Russia. There he found good sequences of undisturbed strata containing the same succession of fossils as in the Silurian, Devonian and Carboniferous of Britain. Above the strata with typical Carboniferous fossils he found thick accumulations of rock, which contained a distinctive set of fossils including some large reptiles. Murchison named these formations the 'Permian System', after the town of Perm on the western edge of the Ural mountains, in 1841.

This work showed beyond doubt that the history of animal life involves a sequence of progressive changes. The Permian fossils provided fresh confirmation that the Tertiary 'age of mammals' had been preceded by a Secondary 'age of reptiles'. This in turn had been preceded by an 'age of fishes' in the Devonian, a point that was clearly demonstrated through the detailed study of fossil fish by Louis Agassiz. Murchison's work showed unequivocally that there had been a still earlier time when invertebrates flourished, before the appearance of any vertebrates. Moreover, when these invertebrate fossils were traced downward through the strata, they simply petered out, giving way to crystalline rocks without any fossils at all. There was every indication that the fossil record could be traced back to a definite beginning.

By the early 1840s, the broad outline of the fossil record had been established. Appropriately enough, the main divisions of this progressive

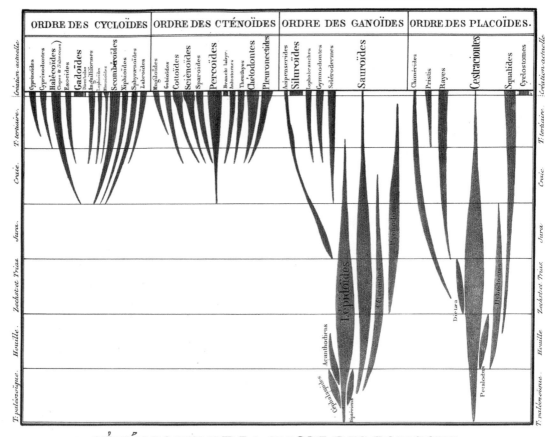

GÉNÉALOGIE DE LA CLASSE DES POISSONS.

sequence were named by William Smith's nephew, John Phillips. In 1841 he proposed that the major eras in Earth history should be designated 'Palaeozoic', 'Mesozoic' and 'Cainozoic', that is to say, the eras of ancient life, middle life and newer life. This classification was soon adopted in place of the earlier categories of Transition, Secondary and Tertiary. The term Tertiary was retained as the name of the main system in the Cainozoic (nowadays spelt Cenozoic).

Given this outline of the history of life, the next question was what it all meant. The usual view was that the changes in species were determined by the changes in the environment. The successive faunas and floras were adapted to the directional changes taking place in the physical Earth. On this view, the appearance of the human species was something of a special case, separate from the rest of creation. Louis Agassiz offered a somewhat different slant on this. He saw the succession of vertebrate fossils as

Agassiz' diagram of fish diversity over time, published in his Researches on Fossil Fish. *Periods of geological time are indicated on the left and Agassiz' revised classification of fishes is set out along the top. The duration and relative abundance of each group is shown, respectively, by the length and width of the line associated with its name.*

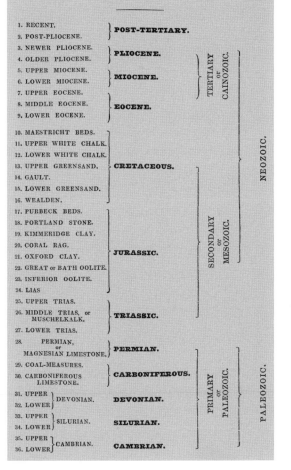

ABRIDGED GENERAL TABLE OF FOSSILIFEROUS STRATA.

1. RECENT.	POST-TERTIARY.		
2. POST-PLIOCENE.			
3. NEWER PLIOCENE.	PLIOCENE.		
4. OLDER PLIOCENE.			
5. UPPER MIOCENE.	MIOCENE.	TERTIARY or CAINOZOIC.	
6. LOWER MIOCENE.			
7. UPPER EOCENE.			
8. MIDDLE EOCENE.	EOCENE.		
9. LOWER EOCENE.			
10. MAESTRICHT BEDS.			NEOZOIC.
11. UPPER WHITE CHALK.			
12. LOWER WHITE CHALK.			
13. UPPER GREENSAND.	CRETACEOUS.		
14. GAULT.			
15. LOWER GREENSAND.			
16. WEALDEN.			
17. PURBECK BEDS.		SECONDARY or MESOZOIC.	
18. PORTLAND STONE.			
19. KIMMERIDGE CLAY.			
20. CORAL RAG.			
21. OXFORD CLAY.	JURASSIC.		
22. GREAT or BATH OOLITE.			
23. INFERIOR OOLITE.			
24. LIAS.			
25. UPPER TRIAS.			
26. MIDDLE TRIAS, or MUSCHELKALK.	TRIASSIC.		
27. LOWER TRIAS.			
28. PERMIAN, or MAGNESIAN LIMESTONE.	PERMIAN.		
29. COAL-MEASURES.			PALEOZOIC.
30. CARBONIFEROUS LIMESTONE.	CARBONIFEROUS.	PRIMARY or PALEOZOIC.	
31. UPPER / 32. LOWER } DEVONIAN	DEVONIAN.		
33. UPPER / 34. LOWER } SILURIAN	SILURIAN.		
35. UPPER / 36. LOWER } CAMBRIAN	CAMBRIAN.		

Table of the local names for the principal British rock formations (LEFT) and their classification into geological periods and eras (RIGHT). Although taken from Lyell's Antiquity of Man *(1863), the materials to draw up such a table were available before 1850.*

the unfolding of a Divine Plan leading directly to the appearance of man, to some extent independently of changes in the environment.

But then, in 1844, the *Vestiges of the Natural History of Creation* was published anonymously by Robert Chambers (1802—71). He argued that the Divine Plan could be realized in an elegant manner if the species of one geological period gave rise to the species of the next period by a process of modified reproduction, in a word by evolution. It seemed obvious to him that some such theory as Lamarck and Saint-Hilaire had proposed was the way to make sense of the new data in natural history. Chambers' book was nineteenth-century pop-science: attractive, successful with a wide audience and quite inaccurate. Although condemned on all sides, the *Vestiges* nevertheless helped push 'the Species Question' to the centre of the stage.

CHAPTER FIVE

THE SPECIES QUESTION

'I do not believe I ever was a fish,' said Tancred in response to a garbled account of evolution from young Lady Constance. But she replied, 'Everything is proved: by geology, you know.' There is only one book that can have inspired this scene in Benjamin Disraeli's novel *Tancred*, published in 1847, and that is Robert Chambers' *Vestiges of the Natural History of Creation*. This scene illustrates the excitement caused by Chambers' popular account of evolution in the *Vestiges*, which was the more exciting for having been published anonymously. Chambers feared that being known as the author of such a book might damage his thriving publishing business.

Certainly he was not thanked by the scientific community for his efforts, but was met with a torrent of abuse. Apart from anything else, scientists were irritated by the book's inaccuracies and complete lack of caution. They were hardly encouraged to take its ideas seriously when they read, for example, that mammals had evolved from birds by way of the duck-billed platypus. So although the *Vestiges* brought 'the species question' to everybody's attention, it can hardly be said to have done much toward providing a satisfactory answer.

As it happens, a sound basis for tackling the species question was already available in Lyell's discussion about species in volume two of the *Principles of Geology*. His treatment of the subject began by outlining Lamarck's views that organisms are in a constant state of flux, with one species continually changing into another. By contrast, Lyell defended the view that species are real units and stable enough to make reliable geological markers. He reviewed the variation that occurs within a species from one place to another and the variation that occurs when species are introduced into new locations. This evidence showed that variation does occur but it appeared to be a strictly limited quantity, with 'indefinite divergence, either in the way of improvement or deterioration, being prevented'.

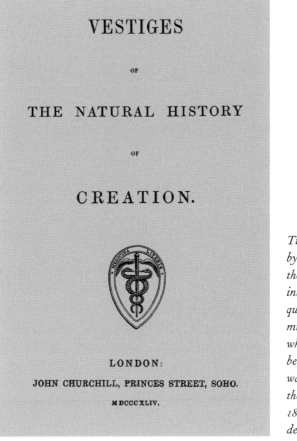

VESTIGES

OF

THE NATURAL HISTORY

OF

CREATION.

LONDON:

JOHN CHURCHILL, PRINCES STREET, SOHO.

MDCCCXLIV.

Title page of the book by Robert Chambers that stirred up popular interest in the species question. There was much speculation as to who the author might be, but his identity was not revealed until the twelfth edition in 1884, after Chambers' death.

Lyell also looked at the breeding experiments of Koelreuter and other hybridizers. Their results indicated that there were natural mechanisms by which species are normally prevented from cross-breeding, and that hybrids obtained experimentally could not perpetuate themselves. 'From the above considerations,' he concluded, 'it appears that species have a real existence in nature, and that each was endowed, at the time of its creation, with the attributes and organisation by which it is now distinguished.' This conclusion about species was much the same as that reached earlier by Linnaeus.

Like Linnaeus before him, Lyell went on to consider the relation between a species and the environment to which it is adapted. The existence of 'distinct botanical and zoological provinces' was an important factor here, he decided. This showed that a particular assemblage of species is restricted to a particular region of the Earth's surface. In view of this, it certainly looked as if each species depended on local environmental conditions and also on the other species with which it interacted in the economy of nature.

At the same time, many species, especially among plants, had the ability to spread far and wide. Their range must then be limited either by their ecological requirements or by some physical barrier. From the work of botanical explorers, one could even predict where barriers to plant distribution might be found in future exploration. In the case of Australia, naturalists had described 'two principal foci of Australian vegetation' at opposite ends of the continent, on the east and west coasts. Lyell predicted, quite correctly, that these extremities would prove to be separated by some physical barrier, 'such as a great marsh, or a desert, or a lofty mountain range'.

Hence the evidence of geographical distribution showed that a given species might expand its range as far as was permitted by physical barriers, suitable conditions and interactions with other species. Equally, Lyell felt, these same factors might cause its distribution to shrink and even to dwindle away to nothing. Living organisms were evidently subjected to a constant 'struggle for existence', in which success led to an expanding range and failure meant a shrinking range. Lamarck had been unable to envisage any natural mechanism for extinction. But Lyell saw clearly that the very fact of adaptation to one set of biological and physical conditions meant that a species was threatened with extinction if those conditions were changed.

Lyell believed that conditions were always changing because the geological agents that he had discussed were continually remodelling the surface of the Earth. The resulting changes in factors like local climate or barriers to dispersal would cause one species after another to dwindle away to nothing. Given continual changes in the course of geological ages, the continual extinction of species was to be expected on ecological grounds. In Lyell's words, 'the successive extinction of animals and plants may be part of the constant and regular course of nature'. If this were so, he realized that people would 'naturally enquire whether there are any means provided for the repair of these losses?'

To this question, Lyell had no satisfactory answer. He hinted that 'new animals and plants are created from time to time' to replace those that have become extinct. And he left it to be inferred that this might be as much part of the regular course of nature as extinction. However, in contrast to extinction, Lyell could not point to any known causes that would be likely to result in the production of new species. He could only suggest that the introduction of new species might be so rare that it was unlikely ever to be observed. This argument bore an embarrassing resemblance to those used by his opponents in defence of sudden revolutions.

Nevertheless, Lyell's discussion was significant because it altered the terms of the debate about species. It effectively set aside the false alternative of extinction versus evolution, which had marked the debate between Cuvier and Lamarck. Instead, extinction was seen as a routine event that could be explained in terms of ongoing geological and ecological processes. This raised at least the possibility that the origin of species might eventually be understood in similar terms. Moreover, Lyell introduced many of the right ingredients into the discussion, such as variability, geographical distribution and the struggle for existence. All this suggested that there must be close links between the problems posed by the history of life and those posed by adaptation and by diversity.

The modern reader may be puzzled by the fact that Lyell writes of species being created, and at the same time implies that species might originate in the ordinary course of nature. Nowadays the term 'creation' tends to be used synonymously with 'special creation', meaning the separate creation of each species by the direct power of God. When the species question was being debated between 1830 and 1860, the idea of creation certainly meant that the natural world had a divine origin. But within this overall idea, it was quite possible to consider how far the origin of species might be due to natural causes. One could ask whether God worked through natural causes, often termed intermediate or secondary causes, or whether a more direct expression of divine power was called for to explain the origin of species.

A notable contribution to this line of discussion was made by Sir John Herschel, who was an astronomer. His scientific interests were broad, which was not unusual in those days, and he wrote a popular book on the nature of science. Entitled *A Preliminary Discourse on the Study of Natural Philosophy*, this book appeared in 1831 and was a well-deserved success. It was from this distinguished background that Herschel (1792—1871) was able to comment on the question of species.

In the *Preliminary Discourse*, Herschel carefully considered what good science should be like. He recognized that scientific theories or laws generally serve two functions: they describe the way the natural world is and they explain how it comes to be that way. First of all the world may be reduced to order by 'empirical laws', which describe the observed regularities but do not explain them. Then there are 'causal laws', which seek to explain the observed regularities in terms of cause and effect relationships. In forming their hypotheses, scientists should always be working towards this second 'higher' kind of law. The way to do this was to search for true causes (*verae causae*), which are those derived from our experience.

After publication of the *Preliminary Discourse*, Herschel was engaged in the task of mapping the southern skies from the Cape of Good Hope. From there he wrote to Lyell in 1836 to endorse the stand taken in the *Principles of Geology* as a good example of the search for true causes. In this context Herschel commented shrewdly on 'that mystery of mysteries the replacement of extinct species by others'. In his opinion, the origin of species should be due to natural causes just as much as their extinction. To think otherwise was to hold 'an inadequate conception of the Creator' since 'in this, as in all his other works we are led by all analogy to suppose that he operates through a series of intermediate causes'. Consequently, if we could ever observe the origin of new species, it 'would be found to be a natural in contradistinction to a miraculous process'.

But Herschel immediately added that 'we perceive no indications of any process actually in progress which is likely to issue in such a result.' This was the point emphasized by those opposed to a natural origin for species, such as William Whewell at Cambridge, who was another scientist with broad interests and competence. Whewell (1794–1866) was sufficiently familiar with geology to review Lyell's *Principles*, and he also wrote on the history and philosophy of science. He developed a sophisticated understanding of science that was similar to Herschel's.

However, Whewell differed from Herschel in thinking that, in the case of species, the combination of evidence required reference to unknown causes. One reason for this opinion was Lyell's obvious failure to incorporate the introduction of new species within his system of uninterrupted causes stretching back into the past. Another reason was the marvellous adaptation of species to their way of life. No one had managed to come up with a natural cause for adaptation, and it still seemed fitting to attribute it to divine design. Although science had shown that cause and effect relationships were the general rule, Whewell argued that there were no causes within our experience that could produce well adapted species.

This was how the matter stood when Charles Darwin (1809–82) came upon the scene. His career as a naturalist had been arrived at by a round-about route. Science was not then recognized as part of the educational curriculum in Britain, and Charles' father had sent him from school to Edinburgh university to train as a doctor. When this did not work out, he was despatched to Cambridge university with a view to becoming a clergyman, but this was not a success either. 'During the three years which I spent at Cambridge my time was wasted,' Darwin later recalled, 'as far as the academical studies were concerned, as completely as at Edinburgh and at school.'

However, it was at Edinburgh that Darwin developed a keen interest in natural history and first heard about the debates between Cuvier, Lamarck and Geoffroy Saint-Hilaire. Darwin brought these interests with him to Cambridge and was soon introduced to the scientific community there. Through his cousin, William Darwin Fox, he was introduced to John Henslow, the professor of botany. Henslow was an able scientist, who lectured on botany, conducted botanical outings and held open house for his students once a week. At these weekly scientific evenings, Darwin became well acquainted with some of the leading scientists of his day, including both Sedgwick and Whewell.

Men like Henslow, Sedgwick and Whewell were part of a growing network of intelligent men who cared deeply about the welfare of science. This they promoted by lecturing in their speciality, by doing research and by participating in the meetings of scientific societies. These societies helped to provide a structure for the scientific community, within which someone like Darwin could be encouraged and nurtured. For instance, Henslow persuaded Darwin to take up the study of geology and got Sedgwick to take him on a geological field trip to Wales. It is evident that his elders detected a vital spark of scientific talent in young Darwin, and they went to some lengths to encourage and guide that talent.

It is interesting to see how the scientific network operated when a certain Captain Fitzroy wished to take a naturalist and gentleman companion with him on a voyage of surveying round the world. Fitzroy made his request to the Hydrographer of the Navy, Captain Beaufort, who asked the mathematician Peacock at Cambridge to recommend someone. Peacock sought the advice of his colleague Henslow, who suggested one of his favourite undergraduates.

Under the command of Captain Fitzroy, HMS *Beagle* stood out to sea from Plymouth Sound in December 1831 and did not return until October 1836. During this voyage of almost five years duration, Darwin proved to be an exceptionally fit and energetic young man. He made extensive collections of animals, plants, fossils and rocks from the places that he visited. He worked continuously at his geology, and had turned himself into a competent and original geologist by the time he returned. He was also brought face to face with many facts about the distribution of living things in space and time, which were not so readily interpreted as the geological phenomena.

When he returned late in 1836, Darwin was welcomed as a vigorous new member of the scientific community in England. He became a firm friend of Lyell, having been convinced of 'the wonderful superiority of Lyell's manner of treating geology' on his voyage. He was encouraged to join the Geological Society, to present scientific papers for publication and to write a popular account of his voyage. With this friendly support, he set about the scientific tasks arising from the *Beagle* voyage in a thoroughly professional manner. It was at this period, when Darwin was sorting out the material from the *Beagle*, that the species question was being debated by Herschell, Lyell and Whewell.

Early in 1837, Darwin's attention was also drawn to this question by news he received from the experts who had taken a look at his *Beagle* collections. One matter of particular interest was what the ornithologist, John Gould, had to say about Darwin's bird specimens from the Galápagos Islands. Many of the specimens apparently belonged to species found only on the Galápagos archipelago. In turn, several of these species belonged to genera confined to the South American mainland and not found widely around the world. Facts like these, Darwin perceived, could be explained if birds from the mainland had been swept out to the islands, where their descendants had become modified gradually and given rise to new species.

Darwin was soon committed to the theory that new species must arise by evolution from pre-existing species. This conviction is reflected in some telling marginal comments in his copy of the fifth edition of Lyell's *Principles*, volume two of which appeared in March 1837. Beside Lyell's claim about 'indefinite divergence ... being prevented', he wrote 'if this were true *adios* theory'. And in July 1837, Darwin opened

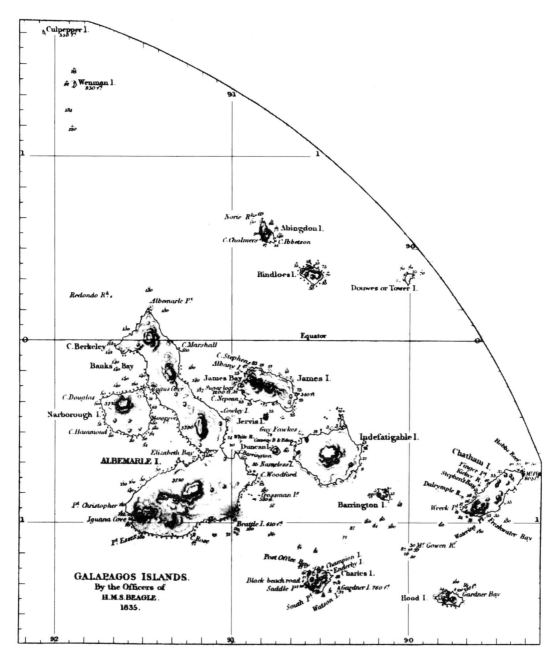

Chart of the Galápagos Islands, prepared by the officers of the Beagle *in 1835, and published in the official* Narrative of the voyage.

his first notebook on the 'Transmutation of Species', a term for evolution used by most authors at that time. As a heading for the notebook, he chose the title 'Zoonomia', in conscious acknowledgement of the work done by his grandfather, Erasmus Darwin, in opening up the subject.

In this first species notebook, Darwin explored the possibilities for change in normal reproduction and variation, and he privately debated

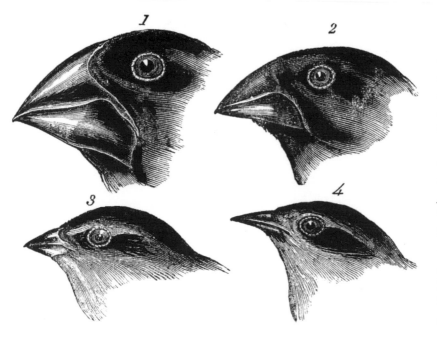

The heads of four different finches found on the Galápagos Islands, illustrated in Darwin's account of the Beagle voyage. These four were recognized and named as separate species by John Gould. 1, Geospiza magnirostris; 2, Geospiza fortis; 3, Camarhynchus parvulus; 4, Certhidia olivacea (see colour plate 9).

the nature of species with Cuvier, Lamarck and Lyell. He also satisfied himself that the evolution of one species into another could explain some important observations he had made in South America. It would explain why extinct forms he had found as fossils seemed to be giant relatives of living forms in the same area. For example, he had found large fossil armadillos and sloths in South America, where similar living species are also found.

Evolution would also explain the well known phenomenon of 'representative species', which is the tendency for a succession of similar species to replace one another across a large continent. In the case of the South American ostrich, of the genus *Rhea*, Darwin had been the first naturalist to collect the representative species of the south, although it was only after the travellers had eaten one for dinner that Darwin realized its importance, and hastily packed its remains off to London. John Gould described this southern form as a species new to science and duly named it after Darwin.

Darwin's notes also discussed how 'my theory' was applicable to other matters such as the pattern of classification. In a world continually modified by geological processes, some species will vary in a way that adapts

A page from Darwin's Notebook of 1837, showing his tree diagram of evolution. 1 is the ancestral species. Evolutionary branches that have become extinct end simply but those that have given rise to surviving species end with a cross-line. These still-surviving species fall into four groups, A, B, C and D, each of which forms a genus.

them to the new circumstances and so form new species. Others will fail to adapt to the changes and become extinct, leaving gaps between those remaining. This produces the pattern of species that we find today, with big gaps between some species and genera but not between others. In evolutionary terms, therefore, 'organised beings represent a tree, *irregularly branched*'. Toward the end of 1837 Darwin noted that 'my theory' must lead to a search for causes: 'with belief of transmutation and geographical grouping we are led to endeavour to discover causes of changes — the manner of adaptation.'

This was the crucial problem for Darwin. Species might change from one into another but the problem was how they change so as to become adapted to new conditions, for organisms are often beautifully adapted to their way of life, and Darwin was much struck with these adaptations. He was also familiar with the argument from design, which held that adaptation was inexplicable in terms of known natural causes. And he was surrounded by senior colleagues, such as Sedgwick and Whewell, who were keen exponents of design. Therefore Darwin knew that, to make a convincing case for evolution, he had to find the natural causes linking modification and adaptation, to discover 'the manner of adaptation'.

Early in 1838, Darwin opened his second species notebook, in which he gave more attention to this problem. In order to solve it, he needed information about individual variations, how they are inherited and how they are related to adaptation. For this he turned to the world of the animal and plant breeder, with which he was familiar from childhood. He noted that the art of successful breeding involves picking out suitable variations that chance to occur and that such variations are usually small ones. The breeder does not make use of monstrosities or 'sports' in his work. From this domestic world, Darwin began to develop an analogy with the natural world.

The idea that natural checks might act selectively on the members of a species was already in circulation among breeders and naturalists. Sir John Sebright, the famous pigeon breeder, declared that 'a severe winter, or scarcity of food, by destroying the weak and the unhealthy, has all the good effects of the most skilful selection'. Passages such as this, containing an idea of natural selection, are to be found in a number of other writers. But they viewed this as a providential mechanism for keeping species and varieties true to type. Darwin's notebooks show that he remained uncertain about the analogy between artificial and natural selection until he read Thomas Malthus' *Essay on the Principle of Population* in September 1838.

One sentence in particular caught Darwin's eye. It is where Malthus said that the human 'population, when unchecked, goes on doubling itself every twenty-five years, or increases in a geometrical ratio'. Given a limited food supply, this exponential rate of increase, as we would now call it, must lead to a desperate struggle for existence among members of the population. This quantitative statement of human population growth enabled Darwin to see just how inexorable and powerful a force the struggle for existence must be in populations of animals and plants. For most of these have the potential to increase much more rapidly than the human population.

Darwin immediately saw that, under these circumstances, any variation that gave an individual some advantage over its fellows would tend to be preserved, and any disadvantageous variations would tend to be eliminated. The struggle for existence would thus exert a strong selective effect, analogous to the breeder picking out desirable variations. So populations of organisms would gradually become modified in ways that adapt them to their environments. By spelling out the exponential rate of population growth, Malthus had made it easy for Darwin to see that the struggle for existence must be a powerful selective force acting on natural populations.

'Here, then, I had at last got a theory by which to work', Darwin wrote later. Having discovered a cause of adaptive change, he promptly set about turning it into a fully worked out theory of evolution. At the end of 1838, Darwin's reading turned to Herschel's *Preliminary Discourse*, which he had first read as an undergraduate, and to Whewell's work. By following their prescriptions on scientific method, his new and controversial ideas would be put forward in a way that professional scientists would respect.

The concept of natural selection clearly met Herschel's requirement for a true cause based on experience. The variability of organisms and the efficacy of selection were known from the experience of breeders, and from this one could infer the selective action of the struggle for existence in natural populations. Darwin decided to make full use of this comparison. He not only read all the leading treatises on breeding but also took up pigeon breeding, a fancy that ran in the family.

As he developed his theory, Darwin tried applying it to a steadily wider range of topics. For Whewell emphasized that one could recognize a good hypothesis by its ability to connect together a wide range of facts that hitherto had seemed unconnected. Darwin took up this approach, arguing that evolution by natural selection made sense of a whole host of facts in previously unconnected areas of study. In this way, the basic structure of his argument became established. He moved from our experience of domestic selection to the struggle for existence and natural selection in the wild. Then he went on to show that a wide range of phenomena could be explained by this theory of evolution through natural selection.

This case for evolution was first sketched out in 1842, and was enlarged into an *Essay* of some 200 pages in 1844. At this stage, Darwin confided in

Joseph Hooker, who was an able botanist and was rapidly becoming a close friend. Hooker read the *Essay* but was not convinced, though he remained sympathetic. So, far from proceeding straight to publication, the *Essay* lay untouched in Darwin's desk for several years. Darwin was well aware that his theory challenged cherished scientific assumptions, such as that of intrinsically limited variability, and that many people would find it repugnant on religious grounds as well. Therefore the theory would carry conviction only if it were supported by a mass of detailed evidence accumulated over a long period of study.

In addition, the year that Darwin wrote his *Essay* also saw the publication of the *Vestiges* by Robert Chambers, whose popular approach did not help the cause of evolution. The furore over the *Vestiges* impressed Darwin with the need to produce a work in an entirely different class. However, there were a few scientists who viewed the *Vestiges* more sympathetically. One of these was the British anatomist, Richard Owen, who wrote a friendly letter to the unknown author saying that he had read the book with 'pleasure and profit'. He went on to say that the discovery of natural causes for producing new species would be well received by his fellow scientists.

For his own part, Owen worked on comparative anatomy in the tradition of Cuvier, whom he had met in 1830. He began to apply Cuverian methods to the reconstruction of fossil animals, and it was he who analysed the extinct mammals that Darwin brought back from South America. It was also Owen who decided that the fossils *Iguanodon* and *Megalosaurus* represented a distinct order of reptiles. For this new order he coined the term *Dinosauria* (Greek for 'terrible lizard') and so gave a name to what were to become the most famous of all extinct animals.

During the early 1840s Owen began to develop some important ideas from his work in comparative anatomy. While he endorsed Cuvier's view of organisms as units of adaptive mechanism, he also found that Geoffroy Saint-Hilaire's unity of plan was a valid idea, as long as comparisons were kept within one major group, such as the vertebrates. Owen's meticulous work showed that exactly the same bones were found in the skeletons of mammals specialized for different ways of life. These same bones could also be identified in the skeletons of other classes of vertebrates. For example, there was a one-to-one correspondence between the bones in the foreleg of a lizard, the foreleg of a mole, the flipper of a dugong and the wing of a bat.

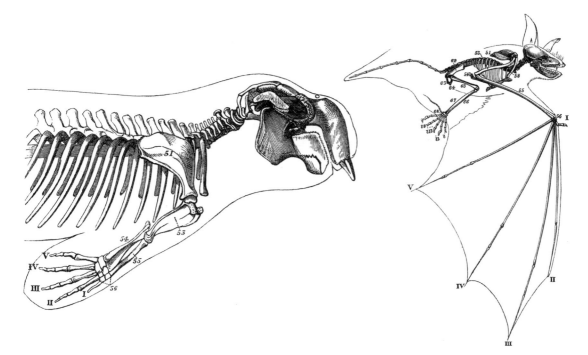

Two drawings from Owen's book On the Nature of Limbs *(1849) to show the homologies of the bones of the forelimb. LEFT, the skeleton of a dugong. RIGHT, the skeleton of a bat. Homologous bones are indicated by the same number throughout the book: 51, scapula; 53, humerus; 54, radius; 55, ulna; 56, carpels. The digits are identified by Roman numerals, with* I *being the thumb.*

Owen used the term 'homology' to denote this correspondence of structure in different animals. He pursued the study of these homologies with great care, and set out his conclusions in *The Archetype and Homologies of the Vertebrate Skeleton*, published in 1848. Owen felt that homologies could not be explained by supposing that homologous parts existed to serve corresponding functions. The existence of exactly the same bones in the forelimbs of many vertebrates could hardly be explained by their special adaptation to different ways of life. Cuverian principles could explain the differences between these limbs but could not account for their underlying similarity.

This difficulty was resolved along the lines laid down by Geoffroy Saint-Hilaire. Owen considered that all animals within each major group represented modifications of a single plan, which he termed the 'archetype'. The vertebrate archetype, for instance, represented the common plan from which all vertebrates, both living and extinct, had been modified according to the adaptive needs of each species. Owen gave the merest hint that this adaptive modification was an actual process in nature, brought about by 'natural laws or secondary causes'. This cautious public comment came at the end of his book, *On the Nature of Limbs*, in 1849, and reflected his private agreement with Chambers' general idea of evolution.

During the 1850s, Owen spent more of his time studying the fossil record. As a result he was able to provide an important revision of the idea of progress in the history of life. That the fossil record was broadly progressive, being dominated successively by invertebrates, fish, reptiles and last of all mammals, was recognized by just about everybody. New fossil evidence convinced Owen that a good case could now be made for gradual progress within each of the vertebrate classes. With regard to the reptiles, the first of the large 'labyrinthodonts' were being unearthed from the Carboniferous system of Germany. These Owen interpreted, rightly enough, as primitive forms that resembled fish and modern amphibians in a number of points. The labyrinthodonts preceded the dinosaurs in time and so gave evidence of gradual progress within the reptiles where there had been none before.

But the new evidence did not fit the simple model of progress in a straight line that Chambers had borrowed to depict evolution. Thus in the case of fish, the great diversity of fossil fish seemed unrelated to an advance toward the reptiles. This was especially so since the great expansion of bony fish took place long after the reptiles had begun their separate development. The true picture seemed to be that many of the ancient animals combined features that were later separated out among distinct modern forms. Cuvier's *Palaeotherium* from the Eocene was a good case in point among the mammals. Within the fossil record of mammals, it was possible to trace out a process of specialization whereby feet, horns and teeth gradually departed from the most general or archetypal condition.

Owen came to the conclusion that this was the usual trend in the fossil record of the vertebrates. The most generalized or archetypal forms were gradually replaced by diverse lines of development, which showed increasing specialization for different ways of life. No longer could the progression of fossil forms be viewed either as a succession of entirely new types, perfectly adapted to successive environments, or as a linear sequence leading to higher forms. Rather, each major group of animals seemed to be introduced in a generalized form; this was then succeeded by more and more refined forms that were adapted to diverse ways of life.

Nor was Owen alone in painting such a picture of the fossil record. In England, William Carpenter (1813–85) emphasized the same trend in his famous *Principles of Physiology*, the third edition of which appeared in 1851.

Drawing mainly on the fossil record of invertebrates, he cited the example of echinoderms. The earliest known echinoderms belonged to the Cystoidea, a group that showed 'a most extraordinary combination of the characters of the remaining groups'. He also drew attention to the parallel between this trend and von Baer's concept of development from general body plan to specialized structures.

In Germany, similar conclusions were drawn by Heinrich Bronn in a prize-winning essay about the history of life. Bronn (1800–85) opened his essay with massive tables of data summarizing the way fossils are distributed through the different rock formations. From these data, he was able to show that the extinction of old species and the introduction of new ones had occurred continuously from the start. While there had been variations in tempo, there had been no occasions when all species had become extinct and been replaced by a completely new set. The living world had approached its present condition gradually, and had shown two main trends. The first was a gradual advance in organization, with later species tending to be more complex than earlier ones. The second was the adaptation of organisms to their environments, which split the development of life into a multitude of separate lines, each advancing in its own way.

Taken together, these two trends produced a history of life resembling a tree, in which new branches were always splitting off in the course of the overall ascent. Bronn was not afraid to tackle the fundamental question of what 'creative force' could have brought about the introduction of all these new species. His use of the phrase 'creative force' does not mean that he believed in special creation, a doctrine that he explicitly denied. Instead, he thought that the creative force must somehow be analogous to the physical forces of gravitation and chemical affinity. The possibility of evolution, as advocated by Lamarck and others, was considered by Bronn but rejected for lack of evidence. He could find no fossil evidence that one species had changed gradually into another.

Meanwhile, the new work on the fossil record was of great interest to Alfred Wallace, who was also thinking about species and their succession. By his own account, Wallace (1823–1913) was an ignorant youth when he first became interested in the species question. He was holding down a job as a school teacher in Leicester when he read the *Vestiges* in 1845, at the age of 22. Wallace was struck by the plausibility of the overall line of argument in the *Vestiges*, and was not put off by its deficiencies.

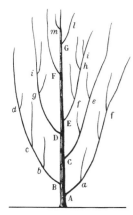

The history of life represented as a tree, a diagram from Bronn's prize-winning essay. The branches represent major groups of organisms, such as the vertebrate classes, and the point at which each branch leaves the trunk corresponds to the first appearance of that group in the fossil record.

'I saw that it really offered no explanation of the process of change of species,' he later recalled. But the view that such change was effected 'through the known laws of reproduction commended itself to me as perfectly satisfactory, and as affording the first step towards a more complete and explanatory theory'.

Wallace differed from other members of the general public who were excited by Chambers' book in that he resolved to do something about it. Together with his close friend Henry Bates, he set off to collect specimens in South America, where they hoped to gather facts relevant to the origin of species. In his travels, Wallace was struck by the way that animal species are geographically distributed over a large area, and this became a life-long interest. For instance, he noted several cases where distinct but closely related species are found on opposite banks of a large river. It was not lost on Wallace that such situations are readily explained in terms of evolution.

Later, Wallace set out on another collecting expedition, this time to the islands of South-East Asia, which he wrote up in *The Malay Archipelago*. There he travelled extensively and once more paid special attention to the geographical distribution of animals. Wallace's notebooks from this period show that he was greatly influenced by Lyell's *Principles of Geology*, which he had read for the first time shortly after he had read the *Vestiges*. It seemed obvious to him that Lyell's position in geology called for some theory like evolution in biology. He considered that it was 'most unphilosophical' to deny that the living world had been produced by a long continued series of still-acting causes, just like the physical Earth.

Like others before him, Wallace accepted Lyell's view of geological causes but combined it with a directional view of the fossil record. His knowledge of fossils came from François Pictet, who had produced a large, four-volume *Treatise on Palaeontology* in 1844–6, with a revised edition appearing a decade later. Like Bronn and Owen, Pictet (1809–72) showed that the approach to the present fauna had been gradual; at no time had all species become extinct and been replaced by completely different species. He also noted that fossil species were formed on the same general plan as living ones but that the older the fossils, the greater the differences from living forms. Fossil species had a limited geological duration that was never interrupted by a period of complete absence from the record.

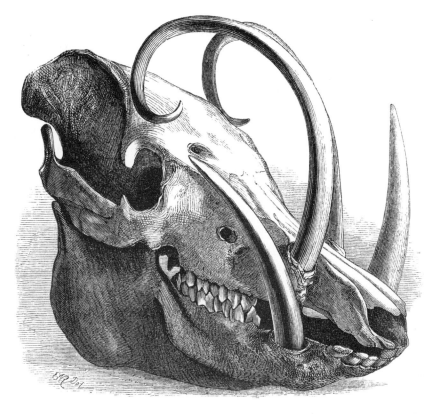

The skull of a male 'babirusa' or pig-deer, a remarkable mammal found only on the island of Sulawesi (Celebes). An illustration from Wallace's account of his travels in The Malay Archipelago. *As he developed his ideas on evolution, Wallace made a special study of such puzzles in geographical distribution.*

SKULL OF BABIRUSA.

Wallace saw that Pictet's results could be readily explained if the later species were related to the earlier ones by evolutionary descent. He thought of writing a book, which would solve the species problem by bringing together the evidence from fossils and from geographical distribution. But in 1854 he was stung into action by a paper by the naturalist, Edward Forbes, which denied progress in the history of life and substituted the idea of 'polarity'. Wallace wrote a reply to show that all the facts fitted quite a different pattern and he put forward his own law to summarize this pattern.

In this paper of 1855, Wallace made public many of the ideas that had been accumulating in his notebooks. He presented his case by comparing the evidence of geography and geology so as to obtain a picture of the distribution of species in space and time. He pointed out that the most closely allied species in the natural system of classification are found in closely adjoining localities. Where countries are separated by some physical barrier, such as a mountain chain or sea, closely allied genera and species are often found on either side of that barrier.

In illustrating this point, Wallace referred to island faunas that are similar to, but not the same as, the fauna of the nearest mainland. He quoted the cases of the Galápagos Islands and Saint Helena, which he knew from Darwin's popular account of the *Beagle* voyage.

Turning to the evidence of geology, Wallace followed Pictet in describing the gradual and progressive changes in the fossil record. He summarized these trends by saying that the distribution of the living world in time 'is very similar to its present distribution in space'. Thus within a given group, species that occur close together in time are more similar than those widely separated in time. From the combined facts of geography and geology, it was evident that species that resembled each other closely were to be found near each other in space and in time. And so Wallace summarized the appearance of new species in this law: 'Every species has come into existence coincident both in space and time with a pre-existing closely allied species.'

This way of stating his views was what Herschel called an empirical law, a description that summarizes the way things are but does not try to explain them. Also in keeping with Herschel's approach, Wallace employed the language of creation but left no doubt that he envisaged a natural mechanism for the origin of species. The living world of today, he explained, 'is clearly derived by a natural process of gradual extinction and creation of species from that of the latest geological periods'. And although Wallace did not say so, the natural process that he had in mind was evolution. This powerful essay implicitly reopened the case for evolution, yet it met with little public response.

However, in private discussion it had a significant impact, especially on Lyell, who was inspired to open his own species notebook. In his notes on the species question, Lyell pondered the shift in opinion that had taken place in the quarter of a century since he had tackled species in the *Principles*. As well as considering Wallace's argument point by point, Lyell also noted down the private discussions that he had with other people on the topic.

These notes confirm that enough progress had been made for naturalists to feel that the species question was no longer beyond the reach of science. While the difficulties in the way of any theory of evolution still seemed formidable, there were steadily fewer naturalists left who cared to defend the fixity of species. In April 1856, Lyell concluded that 'the belief in species as permanent, fixed & invariable, & as

Plate 9 Specimens of Darwin's finches from the Galápagos islands, which were collected by Darwin himself on the Beagle *voyage. These three individuals belong to three separate species, distinguished by the size of their beaks, in the genus* Geospiza: *ABOVE,* G. fortis; *CENTRE,* G. magnirostris; *BELOW,* G. fuligonosa.

Plate 10 The leaves, flowers and fruit of the tree, Eucalyptus urnigera, from Hooker's Flora of Tasmania published in 1860. He noted that the species of Eucalyptus are confined to Australia, and concluded that the major features of plant distribution could be explained by the theory of evolution.

Plate 11 Examples of mimicry among butterflies, described by Bates in 1862. In each pair of butterflies shown, the lower one (4a, 6a, 7a, 8a) is a species of the family Heliconiidae, which are typically brightly coloured and distasteful. And the upper one (4, 6, 7, 8) is a species of Leptalis, family Pieridae, the typical appearance of which is shown by the solitary Leptalis nehemia (5). Bates concluded that the Pierids are mimics of the Heliconiids.

Plate 12B *Examples of discontinuous variation in the colour patterns of adult moths,* from Mendel's Principles of Heredity *by William Bateson. Both species occur in alternative dark and light forms, which are inherited in a simple Mendelian manner with the darker form being dominant. The species* Biston betularia *(3, 4) was later studied as an example of evolutionary change in a natural population.*

Plate 12A *Warning coloration in the caterpillars of British moths. The buff-tip moth,* Pygaera bucephela *(ABOVE), and the cinnebar moth,* Callimorpha jacobaeae *(BELOW), both show the conspicuous patterns typical of warning coloration.*

comprehending individuals descended from single pairs or protoplasts is growing fainter'.

On 16 April 1856, Lyell paid Charles Darwin a visit to talk over Wallace's paper. This prompted Darwin to outline his theory of natural selection to Lyell, who saw at once that it would explain Wallace's law on the introduction of new species. He urged Darwin to publish a preliminary account of his theory as soon as possible, and evidently warned him that Wallace might anticipate him if he did not. Darwin reacted uncertainly to this advice for he was in the thick of his species work. He had recently finished a book on the classification of living and fossil barnacles, which had taken up eight years of his life from 1846 to 1854. This had enhanced his reputation and had been important in testing and confirming his ideas on evolution.

In particular, the barnacle work gave him a valuable insight into the nature of variation. As long as each species was regarded as a fixed and permanent entity, little attention was paid to variability. Individuals of a species were thought of a bit like toy soldiers produced from a mould, which are all identical except for occasional variations due to minor disturbances in the casting process. So it was widely believed that variation was in some way accidental or unnatural, in that it was due to disturbances from outside the reproductive system. Hence variability was thought to be comparatively rare and unimportant, and not to affect the essential character of a species.

In his *Essay* of 1844, Darwin had largely adhered to this view, and supposed that variation was triggered from time to time by disturbances such as geological change. However, his long study of barnacles showed him that variability is a normal feature of animal species. This first-hand study taught him, as no amount of reading or reasoning could have done, that plenty of variation occurs in natural populations. Moreover, the variation was found in all parts of the body, and was not limited to trifling structures of little importance in classification.

Wallace was already aware of the extent of variation from his experience as a collector. Having no formal education in biology, he was more content to take things as he found them. And he found lots of variation in the thousands of insects that he had collected in South America and now in South-East Asia. Wallace therefore took it for granted that variation is a natural consequence of the reproductive process, and so

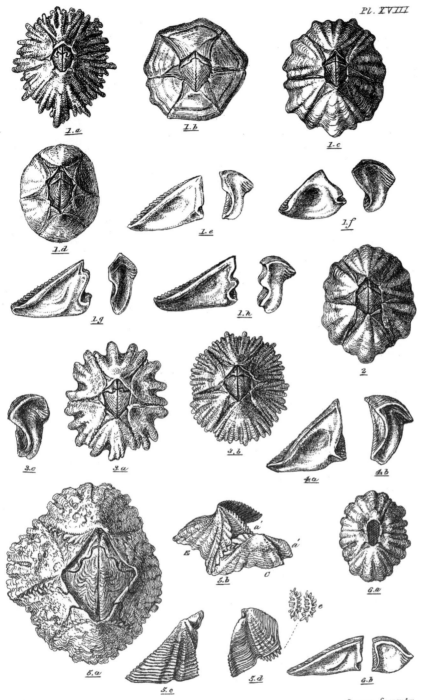

Pl. XVIII

CHTHAMALUS.

George Sowerby

A plate from Darwin's Monograph *on the barnacles, showing species of the genus* Chthamalus. *Variation within a single species is shown in 1a to 1d. The study of barnacles gave Darwin a clear idea of variation in nature.*

needs no special explanation. A plentiful supply of variation was taken as given in his evolutionary thinking.

As Wallace's travels among the islands of South East Asia continued, he struggled with the crucial question of what causes could turn one species into another. He read papers on relevant topics such as varieties and the struggle for existence but, as with Darwin, the final trigger was Malthus. One day early in 1858, while laid low by a bout of malaria, Wallace was thinking over how species could change. His mind turned to Malthus' *Essay on Population*, which he had read long before, and the checks to the increase of human population described there. It occurred to him that such checks must act more effectively on animal populations, which multiply so much more rapidly.

Then the idea suddenly came to him that the individuals removed by these checks must on the whole be those least well adapted to their environment. Given a fresh supply of variation in every generation, it followed that the changes needed to adapt a species to changing conditions would be brought about. Eventually, an entirely new species would arise from the old as the animal's organization was gradually modified in keeping with a changing environment. By the time that his bout of fever had ended, Wallace had thought out the main points of the theory and he wrote it down at once. Over the next few evenings, he composed a short paper and sent it off by the first post to Darwin, whom he knew to be working on the species problem.

Darwin was stunned. Following Lyell's visit in April 1856, he had tried to write up a preliminary account of his theory but had soon abandoned the attempt as hopeless. Instead he began to write a full-length book to be entitled *Natural Selection*. He was over half way through *Natural Selection*, with 11 chapters completed, when Wallace's short paper arrived in June 1858. This paper set out the case for natural selection as a cause that could change one well-adapted species into another with reasoning that was remarkably similar to Darwin's. 'If Wallace had my MS sketch written out in 1842', Darwin wrote to Lyell, 'he could not have made a better short abstract!'

Lyell and Hooker came to the rescue. They arranged for Wallace's paper to be read at the Linnean Society jointly with a paper from Darwin. The latter's contribution consisted of extracts from the *Essay* of 1844 and from a long letter to his friend Asa Gray, who was a leading botanist in the United States. The joint papers were duly read and appeared in the Society's journal for 1858 (see illustration on page 10). This done, Darwin was galvanized into preparing an abstract of his big book for immediate publication. This grew into the volume that we know as the *Origin of Species*, which was published in November 1859.

The final product was worth the long wait, as Asa Gray recognized when he received his copy. He wrote to Darwin: 'I do not think twenty years too much time to produce such a book in.' All those years since the species notebooks had been well spent in working out the detailed application of the theory as fully as anyone could then have done. Furthermore, by following the recommendations of Herschel and Whewell on scientific method, the whole structure of the book was skillfully addressed to contemporary naturalists. Using the basic plan laid down in 1844, it proposed natural selection as a cause of evolution and then applied this theory to a wide range of recent work. In this way, the *Origin* provided a timely synthesis of the latest findings in many areas and showed how these could be explained and unified by evolution.

Darwin set out the plan of his argument in a short introduction to the *Origin*. There he explained that the study of diversity might well lead a naturalist to conclude that species have arisen by evolution rather than by special creation. But this conclusion would remain unsatisfactory until it could be shown how, in the course of evolution, species became adapted to their circumstances. The challenge posed by adaptation was illustrated by reference to the way a woodpecker is adapted to catch insects under the bark of trees, an example noted by John Ray nearly two centuries before. Hence the right point to start was to demonstrate a cause of evolutionary change that could produce well adapted species. As to this cause, Darwin wrote: 'I am convinced that Natural Selection has been the main but not exclusive means of modification.'

Accordingly, he plunged straight into a forty-page account of domestic breeding to provide a basis for his discussion of natural selection. In the case of pigeons, he showed that domestic breeds differ from one another so much that they would be classed as distinct species, or even genera, if found in nature. Yet all these breeds are almost certainly descended from a single ancestral species, the rock pigeon. The means of producing such diverse forms had been the careful selection by fanciers of suitable variations that chanced to occur. Similar variation also occurs in nature, as Darwin showed by a number of animal and plant examples,

A striking example of a domestic breed of pigeon: the English Pouter, from Darwin's Variation of Animals and Plants under Domestication. *Darwin looked to the work of pigeon breeders for evidence about variation and selection, and used this material effectively in the* Origin.

and it is not confined to unimportant structures, as many naturalists had supposed.

He then went on to explain how the struggle for existence is the force in nature that corresponds to the selective breeder in the domestic world. The selective action of this struggle follows from the tendency of natural populations to increase at an exponential rate. This was 'the doctrine of Malthus applied with manifold force to the whole animal and vegetable kingdoms'. That populations can in fact increase exponentially was shown by the cases of dramatic rise in the numbers of cattle, horses and sheep introduced into America and Australia. Yet it was obvious that the populations of wild species cannot all increase significantly 'for the world would not hold them'. Common observation tells us that the numbers of most species remain more or less constant from year to year.

So normally many more individuals are born than can possibly survive, and the members of a population must compete with one another for limited resources. Given a degree of variation among individuals, it follows that some will be in a position to do better than others in this struggle for existence. These better-adapted individuals 'will have the best chance of being preserved in the struggle for life; and from the strong principle of inheritance they will tend to produce offspring similarly characterised'. Long-term change is thus effected, not merely through variation, but through natural selection steadily accumulating those variations that prove advantageous.

When natural selection continues to act over a long time, it will lead to a divergence of structure among the descendants of each successful species. For the more diversified the descendants of a species become, the more they will be able to fill new ecological opportunities and so to increase in numbers. Also, 'more living beings can be supported on the same area the more they diverge in structure, habits, and constitution, of which we see proof by looking at the inhabitants of any small spot'. As the descendents of a successful species diverge and become specialized for different ways of life, so a number of related species will be formed. Continuing divergence, combined with extinction of the less improved forms, will then produce a branching pattern of species and genera.

Having shown how natural selection could be a cause of evolutionary change and adaptation, Darwin was ready to put this theory to the test. Whether it really was able to explain the broad problems of biology had to be 'judged by the general tenour and balance of evidence given in the following chapters'. One of the attractive features of the *Origin* was that Darwin did not pretend that his theory was without difficulties. On the contrary, he made a point of acknowledging these difficulties, and trying to deal with them, before moving on to more positive evidence.

For example, he felt that the most serious difficulty was 'the imperfection of the geological record'. If the species now inhabiting the world have evolved from other species that have become extinct, then we might expect numerous transitional forms to be preserved as fossils. But, as Darwin frankly admitted, 'Geology assuredly does not reveal any such finely graduated organic chain; and this, perhaps, is the most obvious and gravest objection which can be urged against my theory.' In response, he followed Lyell in arguing that rock formations do not represent continuous deposits and that fossilization is a rare event. Furthermore, new species tend to arise as local varieties and so the transitional stages were even less likely to be preserved as fossils. For these reasons, a finely graduated series of fossil forms was not to be expected.

The skeleton of Mylodon, *a fossilized, giant ground-sloth that Darwin found in South America. This reconstruction by Owen showed how it resembles the living sloths, which are confined to South America.*

Apart from such difficulties, the theory of evolution by natural selection made new and good sense of a whole range of recent work. The fossil record itself showed species had changed in the manner that the theory requires, with new species coming in slowly and successively. And the theory would explain why the more ancient a form is, the more it usually differs from living forms, in the way that Owen and others had discerned. Continuing divergence through natural selection would account for the way that many ancient and extinct forms tend to combine features that are separated out among more specialized modern forms. 'For the more ancient a form is', Darwin explained, 'the more nearly it will be related to, and consequently resemble, the common progenitor of groups, since become widely divergent.'

Similarly, the most striking facts of geographical distribution could be explained in terms of evolution and migration. A successful species will first evolve in one locality, and then spread out and diversify, giving rise to an array of closely related or representative species over a large area. This ongoing process would explain why the genera and families of organisms are often localized, and why great barriers to migration mark off separate zoological and botanical provinces. It would explain the bond of affinity that links representative species living in different localities on the one continent. And it would explain why this bond links the living species with the extinct species that formerly inhabited the same continent.

A clear example of these trends was that of oceanic islands such as the Galápagos. These have a fauna and flora similar to, but not the same as, that of the nearest mainland, and this makes sense in terms of migration and evolution. For some of the more mobile species from the mainland would eventually reach newly formed islands, and once there they would evolve into new species better suited to the local conditions. These processes would explain why oceanic islands are inhabited by relatively few species, most of which are found nowhere else. They would also explain why amphibians and terrestrial mammals should be absent from oceanic islands even though the most isolated islands possess their own species of bats.

Again, evolution by natural selection would explain the pattern of life seen in the hierarchy of classification, with group subordinate to group. Divergent evolution would be expected to result in just such a branching pattern. Hence 'community of descent is the hidden bond which naturalists have been unconsciously seeking' in their search for a natural system of classification. In the same way, evolution would make plain sense of homology by putting a common ancestor in place of a mysterious archetype. The presence of homologous bones in the foreleg of a horse, the wing of a bat and the flipper of a seal was due to inheritance from a common ancestor. 'On my theory,' Darwin said simply, 'unity of type is explained by unity of descent.'

Finally, where species of animals are placed in the same class on the basis of adult structure, the similarity of their embryos is elegantly explained by evolution. As a rule, the detailed differences between species in a class tend to appear relatively late in development. During evolution, natural selection will add more and more such differences to the adults but this process will leave the early embryos relatively untouched. Hence the

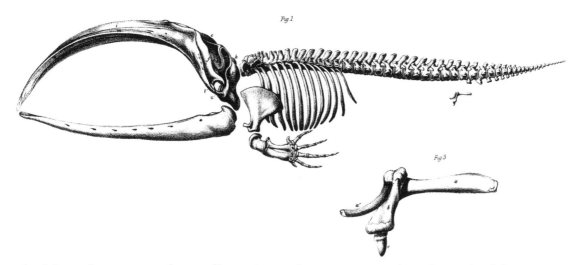

Fig 1

Fig 3

The skeleton of a Greenland Right Whale, first described in 1861. If the Origin *had been illustrated, this would have been a perfect example since it demonstrates homology in the bones of the fore limb (compare with illustration on page 115) as well as rudimentary structures in the hind limb. The rudimentary hind girdle and limb are shown enlarged in the inset (BELOW); the pubis (a'), ischium (a) and femur (b) can just be distinguished as separate bones.*

embryo will remain as a 'picture, more or less obscured, of the common parent-form of each great class of animals'.

Here, Darwin was especially pleased with his ability to explain rudimentary structures, which are often conspicuous in the embryo but partly or completely disappear in the adult. Examples include the foetal teeth in the jaws of baleen whales (see colour plate 8A) and the vestiges of side-toes in the legs of horses. Hitherto these puzzling cases were generally said to have been created 'to complete the scheme of nature'. But in terms of evolution, their origin was simple: they were the now atrophied remnants of parts that had once been fully formed in ancestral species. Rudimentary organs, Darwin explained, 'may be compared with the letters in a word, still retained in the spelling, but become useless in the pronunciation, but which serve as a clue in seeking for its derivation'.

In the last chapter, Darwin noted that 'this whole volume is one long argument', and helpfully pulled the main threads together. The breadth and cogency of that argument gave the *Origin* its power, and make it fascinating to read even today. By bringing together such a wide range of recent work, including much that was at first sight unfavourable to his theory, Darwin made the case for evolution convincing in a way that it had never been before. As well as being up to date, the *Origin* was short enough to give its arguments their full force. Since it was written in a hurry, it did not have massive compilations of data, nor was it encumbered with footnotes. Taken together, the logic of its argument, the breadth of its synthesis and its convenient length gave Darwin's book an overwhelming impact.

CHAPTER SIX

LIFE'S GENEALOGY AND NATURAL SELECTION

'How extremely stupid not to have thought of that!' was Thomas Huxley's reaction when he first read the *Origin of Species*. He was able to read it through while supervising an examination late in November 1859, and he saw at once that Darwin's theory could solve the species problem. Huxley (1825–95) was then a young zoologist who had made his name in comparative anatomy. His career had begun in the previous decade, on board HMS *Rattlesnake*, where he had studied the invertebrates that float near the surface of the sea. He had met Darwin shortly after his return from the voyage of the *Rattlesnake*, and the two men were firm friends by the time the *Origin* was published.

Huxley illustrates the impact made by the *Origin* rather well, because he had little idea of Darwin's views before reading the book. He had come up against the species question in the 1850s, and soon put aside the idea of special creation as advocated by Agassiz and others. Nor had he any time for Owen's abstract archetypes, which he thought were just 'verbal hocus-pocus'. Although Huxley felt that it must be possible to bring the origin of new species within the range of science, he was not at all impressed by Lamarck's theory of evolution or any later modification of it. His opposition to evolution was made clear in a savage review of one of the later editions of the *Vestiges*.

But when the *Origin* appeared, it impressed him at once as a 'noble book' that really got to grips with the available data. Here at last was a theory that shed a flood of new light on the whole vexed question. 'I am prepared to go to the stake, if requisite,' he assured Darwin, in support of the chapters on the geological and geographical distribution of life. Huxley did not offer quite the same support for the chapters on natural selection, although he agreed that 'you have demonstrated a true cause for the production of species'.

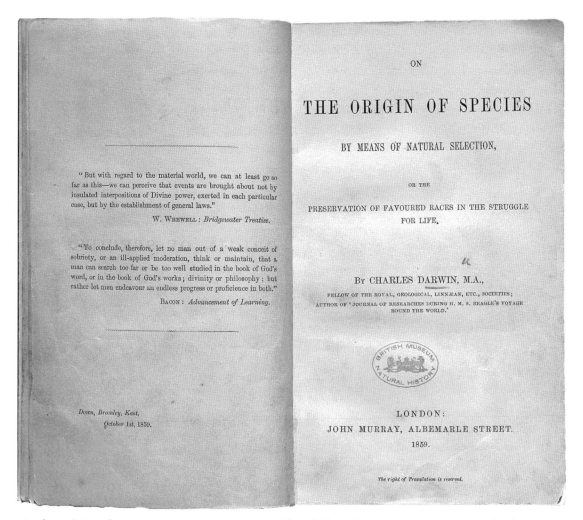

ON

THE ORIGIN OF SPECIES

BY MEANS OF NATURAL SELECTION,

OR THE

PRESERVATION OF FAVOURED RACES IN THE STRUGGLE
FOR LIFE.

By CHARLES DARWIN, M.A.,

FELLOW OF THE ROYAL, GEOLOGICAL, LINNÆAN, ETC., SOCIETIES;
AUTHOR OF 'JOURNAL OF RESEARCHES DURING H. M. S. BEAGLE'S VOYAGE
ROUND THE WORLD.'

LONDON:
JOHN MURRAY, ALBEMARLE STREET.
1859.

The right of Translation is reserved.

"But with regard to the material world, we can at least go so far as this—we can perceive that events are brought about not by insulated interpositions of Divine power, exerted in each particular case, but by the establishment of general laws."

W. Whewell: *Bridgewater Treatise.*

"To conclude, therefore, let no man out of a weak conceit of sobriety, or an ill-applied moderation, think or maintain, that a man can search too far or be too well studied in the book of God's word, or in the book of God's works; divinity or philosophy; but rather let men endeavour an endless progress or proficience in both."

Bacon: *Advancement of Learning.*

Down, Bromley, Kent,
October 1st, 1859.

The first edition of Darwin's Origin of Species, *open at the title page. Note the quotations from Francis Bacon and William Whewell placed opposite the title page.*

Darwin was especially glad to have Huxley's support for his views. As he rushed to put the *Origin* together, Darwin had decided that Lyell, Hooker and Huxley were the men whose assessment of it he would value most. With their great practical experience and keen minds, these three friends would provide the critical test of whether or not his theory was along the right lines. In the event, all three heaped praise on the *Origin* as an exceptional piece of work. They were also convinced by its argument and accepted the idea that new species arise by 'descent with modification' from pre-existing species. However, although they agreed that natural processes could transform one species into another, they were not convinced that natural selection was adequate for the task.

Among the wider audience of naturalists, the *Origin* met with a stormy reception at first. The year 1860 saw many reviews of it appear, most of which were rather hostile. Naturally, the majority of reviewers were used to looking at things quite differently from Darwin, and so were taken aback by his line of argument. But Darwin carried the day, and in a remarkably short time at that. Within ten years most working naturalists in the English-speaking world had accepted the general concept of evolution. When he came to publish the last edition of the *Origin* in 1872, Darwin could comment with some satisfaction: 'Now things are wholly changed, and almost every naturalist admits the great principle of evolution.'

Naturalists were ready to accept evolution because they saw that it spoke to the problems of their daily research. Confronted by the diversity of life, they found problems even at the simplest level since it was often difficult to distinguish between true species and local varieties within a species. And this difficulty increased, instead of getting less, as their studies were extended. This should not have happened if each species had been separately created with a distinct and permanent structure, which was the view that had prevailed since the time of Linnaeus. On the contrary, it should have been possible to distinguish species clearly by criteria such as structural differences and mutual sterility when cross-bred.

One reviewer of the *Origin* in 1860, who acknowledged that species are not as sharply defined as one would expect if they had been specially created, was William Carpenter, the physiologist. 'Naturalists have gone on quite long enough on the doctrine of the "permanence of species",' he wrote. 'Their catalogues are becoming more and more encumbered with these hypothetical "distinct creations".'

In contrast to this, Carpenter stated, the best naturalists recognize that 'no species can be fairly admitted as having a real existence in nature, until its *range of variation* has been determined both *over space* and *through time*'. A study along these lines of the British flowering plants had recently led to a reduction in the number of true species by about a quarter. Here Darwin's work was of the 'highest value' since it provided a new principle that made good sense of this variability. So Carpenter was much inclined to accept Darwin's basic thesis, at least for groups of similar organisms.

There were a few naturalists, of course, who rejected any such line of thought and stuck to the earlier view that each species had been created separately. Two notable examples were Darwin's old teacher, Adam Sedgwick, and the Swiss naturalist, Louis Agassiz, now living in the United States. Faced with the geographical diversity of life,

Agassiz maintained that each species had been specially created in just that locality where it is now found. Each species was designed to fit its allotted locality even in such details as population size. On this principle, Agassiz was obliged to call upon a fresh act of creation even for marked local varieties within a species. To people like Carpenter, it seemed more fitting to suppose that God worked through natural laws when introducing new species. Surely God's power and glory were revealed more clearly in natural laws than in a peppering of miraculous interventions.

As Lyell observed, the very absurdity of Agassiz's position helped to make Darwin's views more acceptable among naturalists. For his own part, Lyell's belief in the separate creation of each species was shaken more than anything else by evidence from plants and animals introduced into countries colonized by Europeans. In certain cases, the introduced species had become naturalized, and some were so successful that they displaced species native to the area. This kind of situation showed clearly that each species is not uniquely designed for its present locality as Agassiz had thought. The fact that, given the opportunity, some species could flourish and even take over in areas other than their native locality was much more consistent with Darwin's views.

These comments from Carpenter and Lyell illustrate the sorts of consideration that persuaded the great majority of naturalists to adopt evolution as a theory that they could work with. The geographical distribution of plants and animals was one field of study where naturalists soon saw that evolution could make sense of their research. Hooker was one of the first to use the new ideas, in his studies of the regional floras of the world. In the 'Introductory Essay' to his *Flora of Australia*, published in 1860, he noted that naturalists who considered species to be immutable creations were unable to agree on how many species of flowering plants there were. However, if Darwin was right, there were bound to be doubtful cases in between local varieties and full species.

To Hooker, the way that species and genera of plants are, as a rule, confined within certain geographical limits also supported Darwin's views (see colour plate 10). This arrangement would be the natural result if the species of each genus had sprung from a common parent and then spread in various directions from a common centre. In fact, the arrangement of plants across the world was just what one would expect if change had been going on indefinitely 'so as gradually to give rise in the course of time to the most widely divergent forms'.

New evidence of this kind was marshalled for animals by Wallace. Some of his best material was gathered from his long stay among the islands of South-East Asia, from which he collected thousands of specimens. He found that the animals characteristic of tropical Asia were replaced by those characteristic of the Australian region as one travelled eastwards. On the whole, this transition was gradual, but Wallace was surprised to discover a relatively sharp change in the middle of the island chain. From the distribution of bird species, he concluded that this dividing line between the two faunas must run between the islands of Bali and Lombok, which are only a few miles apart. His reports on this remarkable pattern of distribution were published in 1859 and 1860.

From the outset, Wallace was keen to explain, not merely describe, the geographical distribution of animal species. And his prediction was that the current distribution of the terrestrial species would be found to reflect the geological history of the world's land masses. He was therefore amazed that the Asian and Australian faunas were not separated by any major physical or climatic barrier.

'Facts such as these,' he wrote, 'can only be explained by a bold acceptance of vast changes in the surface of the Earth.' If there was no barrier now, there must have been one in the past to account for the evolution of such a marked separation between two distinct faunas. From the geography of the continental shelf below the sea, Wallace knew that the two faunas were connected with two distinct continental areas. He supposed that these continents must have risen and fallen; it was left to the twentieth century to show that they had actually moved closer together.

Some years later, in 1876, Wallace put the new outlook together in a major book of two volumes, *The Geographical Distribution of Animals*. Here he showed that the present distribution of animals around the world cannot be 'wholly due to diversities of climate and of vegetation'. In so many cases, countries were similar in climate and physical geography and yet had quite distinct kinds of animals. This situation could only be understood in terms of changes in the past, both in living things and in the structure of the land and oceans. The present distribution of animals was 'the result and outcome of all preceding changes of the earth and of its inhabitants'. Wallace therefore took into account the distribution of fossil animals as well as all the main groups of living terrestrial animals.

This bold new approach enabled him to explain many puzzling cases of distribution, such as that of the camel family. Living species of camel are

Two plates from Wallace's Geographical Distribution of Animals, *illustrating the contrast between animals of the Oriental and Australian regions. LEFT, a selection of animals characteristic of Borneo, including a tarsier* Tarsius bancanus *(top left), a treeshrew* Ptilocercus lowii *(centre left) and a tapir* Tapirus indicus *(centre right). RIGHT, Wallace's selection of animals from New Guinea to contrast with those from Borneo, including a tree kangaroo* Dendrolagus inustis *(top left), a lory* Charmosyna papou *(top right) and a racquet-tailed kingfisher* Tanysiptera galatea *(lower centre).*

A FOREST IN BORNEO, WITH CHARACTERISTIC MAMMALIA.

SCENE IN NEW GUINEA, WITH CHARACTERISTIC ANIMALS.

confined to deserts, from the Sahara to Mongolia, while the closely related llamas are found only in the mountains and deserts of South America. However, camel-like fossils, which seemed to qualify as evolutionary ancestors, were to be found in the Pliocene rocks of North America. Hence ancestors of the camel family occurred 'in a region where they do not now exist, but which is situated so that the now widely separated living forms could easily be derived from it'. In this way, Wallace was highly successful in applying evolutionary theory to the problems of animal distribution around the world.

Darwin greeted this book with 'unbounded admiration', and assured Wallace that it had 'laid a broad and safe foundation for all future work on Distribution'. The diversity of life, both in classification and geography, clearly made sense when seen as the result of evolution influenced by the geological history of the planet. Direct evidence about the genealogy of life was to be found in the fossil record of animals and plants, as Wallace had recognized. The history of life revealed by the study of fossils would therefore provide a crucial test for the theory of evolution. If groups of similar organisms were truly descended from a common ancestor, then it should be possible to trace this ancestry in various ways. So the genealogy of life became the main research topic for several fields of study in the latter part of the nineteenth century.

A typical reaction to the theory in this context was that of François Pictet, author of the respected *Treatise on Palaeontology*. This was the book that had been such a help to Wallace as he had prepared his 1855 paper on the introduction of new species. Reviewing the *Origin* in 1860, Pictet praised it as a comprehensive work that was refreshingly different from 'the ordinary routine'. He was 'quite willing to accept most of the facts and the ideas' set forth in the *Origin*, but only up to a certain point.

It was one thing to accept that individual variation, acted on by natural selection, might give rise to new species of fairly similar form. It was quite another thing to see how major groups of organisms, which differ greatly in their organization, could have originated by the same means. In any case, the fossil record failed to provide any evidence of such gradual transitions. Pictet stated that he could not accept evolution on this larger scale until he saw direct evidence of major changes effected by the means Darwin suggested.

Nevertheless, he acknowledged that Darwin had been able to produce indirect evidence 'whose relevance is real and incontestable'. The theory provided 'admirable explanations' for unity of plan, for rudimentary organs and for the arrangement of species and genera in the natural classification. The theory made equally good sense of the resemblance between fossil species in successive formations and of the parallel sometimes seen between the fossil record and development in the embryo. In the end, Pictet was won over by the power of this indirect evidence. By 1864 he had accepted large-scale evolution, and two years later he wrote a paper on fossil fishes that lent support to evolutionary theory.

In his review, Pictet had explained that Darwin's imagination 'advanced more quickly than mine' in taking the theory of evolution quite so far. This was how Carpenter and many other naturalists also felt. It was not that they rejected the whole idea, but rather that their imaginations needed to be stretched gradually to accommodate evolution. Nothing contributed more to that enlargement of mind than the discovery of fossil forms intermediate between those already known. The fossil record might be too incomplete at a detailed level to record the change from one species to another, as Darwin had claimed. But it was widely agreed that it was complete enough to record a broad outline of the history of life, especially among the vertebrates. It should therefore be able to show whether fossil forms succeeded one another in a way that was consistent with Darwin's theory.

The ink was scarcely dry on the first reviews of the *Origin* before a fossil discovery was made that provided the right kind of evidence. In 1860, a fossil feather was found in the limestone deposits of Solnhofen in Bavaria, and the following year saw the recovery of an almost complete skeleton adorned with feathers from the same deposits. This remarkable find was given the name *Archaeopteryx* (Greek for 'ancient wing') and soon became one of the most famous fossils. The Solnhofen deposits were of Jurassic age, much earlier than any previous record of a fossil bird.

This first specimen of *Archaeopteryx* was purchased by the British Museum, where Richard Owen was now in charge of the Natural History collections. He published an accurate description of this fossil in 1863 and concluded that it was 'unequivocally a bird'. At the same time, he noted that some of its features were found only in the embryonic stages of living birds, and that it showed 'a closer adhesion to the general vertebrate type'. In particular, it possessed a long bony tail with a pair of feathers attached to each vertebra, and the forelimbs had separate digits complete with claws. A further study of the fossil suggested that it also had teeth, which was confirmed when an even more complete specimen was obtained from the Solnhofen quarries in 1877.

When he came to read this description, Huxley saw that the embryonic or generalized features, such as the long tail, separate digits and toothed jaws, were clearly reptilian. *Archaeopteryx* was therefore a transitional form that helped to bridge the gap between birds and reptiles. Moreover, the Solnhofen deposits also contained a small bipedal dinosaur called *Compsognathus*, which was bird-like in certain respects and so could go some way towards closing the gap from the other side. In a popular lecture in 1868, Huxley argued that *Archaeopteryx* and *Compsognathus* provided a test case that gave strong support to the theory of evolution. The existence of a reptile-like bird and a bird-like reptile at an appropriate

A late nineteenth-century restoration of Archaeopteryx, *based on the two specimens then known. The bird's reptilian features, such as toothed jaws and long bony tail, are clearly shown. The inset (BELOW), which shows the pair of feathers attached to each tail vertebra, has been copied from Owen's original description.*

level in the fossil record showed that an evolutionary transition from one class of vertebrates to another was entirely plausible.

Another set of fossil discoveries, in which Huxley took a keen interest, was concerned with the genealogy of the horse. The first of these discoveries was made by Albert Gaudry (1827–1908), who worked on fossil-bearing rocks at Pikermi in Greece, as part of a French expedition to the area in the 1850s. These rocks yielded a rich assortment of fossil mammals, which were of special interest because they belonged to the Miocene and Pliocene periods. These are intermediate in age between the two mammalian faunas studied by Cuvier, those of the superficial gravels (Pleistocene) and of the Parisian quarries (Eocene). Gaudry found that many of these mammalian fossils were intermediate in structure

Bones of the lower forelimb in four genera of hoofed mammals from succesive intervals of the Tertiary period: A, Hyracotherium (Eocene); B, Anchitherium (Miocene); C, Hipparion (Lower Pliocene); D, Equus (the modern horse). This illustration shows the gradual reduction of the side toes over time.

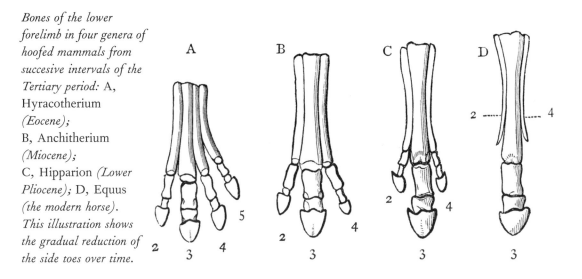

between the species recovered from the earlier and later periods. By far the best way to interpret these intermediate forms, he felt, was in terms of evolution.

In the case of the horse family, living horses and those of the Pleistocene deposits were placed in the genus *Equus*. This is a highly specialized genus of hoofed mammal (ungulate), notable for having the foot reduced to a single elongated toe. By contrast, Cuvier's *Palaeotherium* from the Eocene is a much smaller and more generalized ungulate, with three toes of nearly equal length. At Pikermi, Gaudry found abundant fossils of a less specialized horse-like mammal, *Hipparion*, in which side-toes were clearly present although they were too small to reach the ground. Since *Hipparion* was intermediate between the Pleistocene and Eocene forms both in structure and in geological age, it could represent an evolutionary link between them. So when he published his work in the 1860s, Gaudry used *Hipparion* to draw up a provisional genealogy of modern horses in the form of a branching evolutionary tree.

This genealogy was taken much further by Othniel Marsh (1831–99), who had visited Western Europe from the United States to study the fossil collections. His search for similar fossil sequences in his own country was richly rewarded. By 1874 he had found a series of fossils that showed 'every important intermediate form' between a small unspecialized *Orohippus* in the Eocene and the large specialized *Equus* of today. This series revealed a gradual reduction in the side toes of each succeeding species, as well as corresponding changes in other limb bones and in the teeth. In a lecture in 1877, Marsh likened fossil intermediates such as the horse series and *Archaeopteryx* to 'stepping stones' that led across 'the shallow remnant of the gulf once thought impassable'. There could no longer be any doubt that evolution had occurred; it provided 'a key to the mysteries of past life on the earth'.

A few years later, this genealogy was taken even further back in time by Edward Cope, whose energetic search for new fossils led to his famous rivalry with Marsh. Cope (1840–97) discovered a fossil mammal, which he named *Phenacodus*, in the early Eocene rocks of North America. This had a foot with five digits, each of which had a simple hoof in place of a claw. Its generalized structure and early date meant that it could be close to the ancestor for all the hoofed mammals.

From this beginning, structures such as horns, teeth and feet appeared to have evolved along a regular path, reminiscent of the regular pattern of development in the embryo. Moreover, Cope noted, similar changes had taken place in separate lines of evolution. In the case of feet, these became strengthened by an interlocking of the ankle bones and by a reduction in the number of digits, taking place independently in separate groups. This reduction led to only two toes in camels and deer, and to a single toe in the horse series. So it was clearly possible to find the probable ancestors of modern vertebrate groups, given a rich assemblage of fossils like this. And in some cases it was even possible to trace the step-by-step sequence of evolution.

Meanwhile, the excitement over evolution gave rise to the idea that it should be possible to draw up a complete genealogy of all animals and plants. Where the fossil record was lacking, relevant information could be gleaned by comparing the morphology of living species and especially their embryos. This idea was taken up with great enthusiasm by Ernst Haeckel (1834–1919) in Germany, who wrote a series of popular books starting in the 1860s. In the following decade, these were translated into English under titles such as *The History of Creation* (1868) and *The Evolution of Man* (1879). Several of the words and images that are now commonplace in biology were coined by Haeckel in these books. He illustrated the evolution of life by constructing family trees, often rendered as wonderfully life-like drawings.

Within this overall scheme of things, Haeckel developed the concept of the 'phylum' (Greek for 'stem') to accommodate all the species descended

The fossil mammal Phenacodus *from the Eocene, discovered by Cope. This skeleton is almost complete and shows the distinctive features of the earliest hoofed mammals.*

from a common parent-form. To denote the sequence of their evolution from this parent-form he introduced the term 'phylogeny'. Similarly, he suggested the term 'ontogeny' to denote the development of the individual from fertilized egg to adult within each species. That a certain parallel exists between the development of an individual and the sequence of forms in the fossil record had long been recognized for vertebrates. Darwin adopted this parallel cautiously in the *Origin of Species*, since it could be explained by supposing that ontogeny must reflect phylogeny to some extent.

The idea that an individual passes through the stages of its species' evolution as it develops from egg to adult was adopted boldly by Haeckel. He expressed it by saying that 'ontogeny is a concise and compressed recapitulation of phylogeny'. Thus the human embryo could be said to recapitulate the course of evolution during its nine months in the womb. It begins as a single cell at fertilization, and after few weeks the organization of the foetus is equivalent to that of the simpler vertebrates. Gill slits, which are characteristic of fish, appear for a time in the head region; the backbone also appears and extends well beyond the legs to form a tail, which is characteristic of most vertebrates.

PEDIGREE OF MAN.

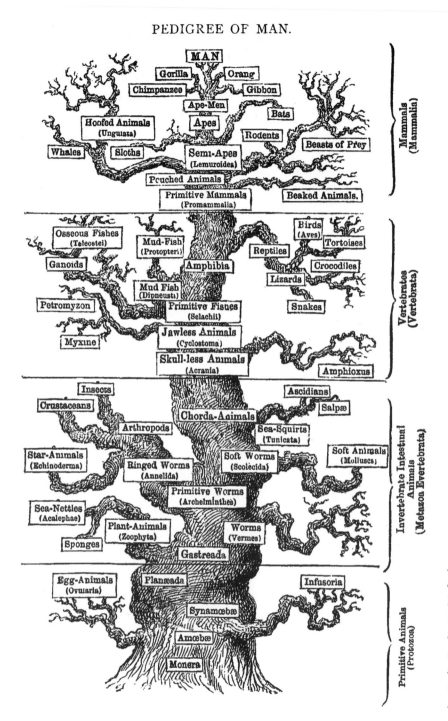

An evolutionary tree of life, from Haeckel's Evolution of Man. *The lifelike character of the drawing tended to conceal the largely speculative nature of the relationships shown.*

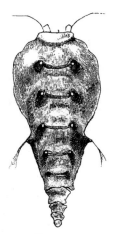

GORILLA

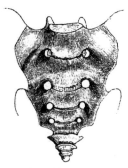

MAN

Rudimentary tail bones present in the skeletons of gorillas and humans. Facts like these gave early support to the theory of evolution.

Haeckel was able to show that at this early stage there is not much to choose between the embryos of bird, dog and human: they all resemble a simplified vertebrate. Only at a later stage do the differences between them make their appearance. Hence the study of embryos provided good evidence for the common ancestry of all vertebrates, including humans. Where the embryological evidence could be compared with the study of fossils or of rudimentary organs, it proved to be especially compelling. For instance, the presence of a long tail in the human embryo is consistent with the fact that human adults possess a set of rudimentary tail-bones. Taken together, these two pieces of evidence strongly suggest that a tailed vertebrate ancestor forms part of human phylogeny.

Haeckel believed that recapitulation could supply an almost infallible key to the evolutionary ancestry of the animal kingdom. Embryology would provide a firm framework of phylogeny, into which the fossil evidence could be fitted as it became available. Where no scientific evidence was available, he used his own creative logic to fill the gap. But it soon became clear that recapitulation did not hold up at the detailed level Haeckel had hoped for, and his theory lost its appeal after the turn of the century. Most zoologists were content to use embryology as evidence for evolution in general, without expecting it to yield detailed information on phylogeny.

In other respects, too, Haeckel got quite carried away. His books turned evolution into an all-embracing philosophy that went far beyond anything endorsed in the *Origin of Species*. Similarly, in England, evolution was taken on board by the fashionable philosopher of the day, Herbert Spencer. He just absorbed Darwin's theory into his philosophical system, as the biological part of a cosmic principle of 'development'. It was he who coined the term 'survival of the fittest' to describe the process of natural selection. In their day, the generalizations of Haeckel and Spencer were accorded great respect and were even welcomed by people like Huxley. When he lectured on *Archaeopteryx* in 1868, Huxley was a firm supporter of Spencer's cosmic evolution, but he withdrew this support later in life.

One point that Haeckel had taken for granted from the start was that humans were included in the evolutionary scheme of things. In addition to the embryological evidence that he brought forward, other lines of evidence gave support to this conclusion in the 1860s. One of these was comparative adult anatomy, and Huxley set this out for all to see when he published *Evidence as to Man's Place in Nature* in 1863. Huxley began by showing that scientists had finally got a clear picture of the primates, the zoological group into which Linnaeus had placed the human species. The primates that resembled humans most were the great apes, he explained, which were large primates without tails. Four different kinds of ape could be clearly distinguished: the gibbons and orang-utans in Asia, and the chimpanzees and gorillas in Africa.

Huxley went on to compare the bodily structure of these apes with that of humans, paying special attention to the skeleton and the brain. He showed that there are no parts of the body unique to humans, not even in the brain — a point on which he had clashed with Richard Owen. Certainly there were differences between an ape and a human but these were actually less than the corresponding differences between an ape and a monkey. 'So far as cerebral structure goes,' Huxley affirmed, '... Man differs less from the Chimpanzee or the Orang, than these do even from the Monkeys.' Comparing other parts of the body told the same story. No absolute structural line of demarcation could be drawn between man and ape any more than between ape and monkey.

In the same year that Huxley's book appeared, Charles Lyell also turned his attention to the human species, with a book on the *Antiquity of Man*. Here he reviewed another line of evidence, that provided by the stone implements made by the people of an earlier time. Such implements had been dug up for centuries, and a few finds were carefully recorded, but they were not seen as having any special significance. The first person to collect them systematically was a French customs official by the name of Boucher de Perthes (1788–1868). He took to searching for fossils in gravel pits near Abbeville on the river Somme, and in 1838 found a stone axe that was unmistakably man-made. As he continued to explore, he was excited to find more stone implements, and associated with them the bones of mammoths, woolly rhinoceroses and other extinct species.

When Boucher de Perthes published a full account of this work in 1846, few people were willing to believe him at first. But in 1854 similar stone axes were described that had been found at St Acheul, near Amiens, further up the Somme valley. The interest of these two sites led to their being visited by many geologists, including Lyell himself in the spring of 1859.

A flint hand-axe collected near St. Acheul in France. This illustration from Lyell's Antiquity of Man *shows both a face and a side view of the same implement.*

This first-hand experience combined with the assessments of other experts to convince Lyell that the finds were genuine. There are, he wrote in the *Antiquity of Man*, a 'wonderful number of flint tools, of a very antique type, which . . . occur in undisturbed strata, associated with the bones of extinct quadrupeds'. There could no longer be any doubt that humans had existed on Earth at a much earlier period than geologists had previously thought.

A further line of evidence that pointed in the same direction was the occurrence of human bones in a fossil state. Lyell and Huxley put their skills together to assess these recent finds, one of which was to become particularly famous. It consisted of fragments of a human skeleton uncovered in 1856 by quarrymen excavating a cave in the Neander valley (tal) of Germany. Following their formal description in 1858, these remains became known as Neanderthal man and formed a topic of controversy and speculation for years. The circumstances of its discovery made it impossible to date this find accurately, but Lyell agreed that it was indeed old and probably belonged with the extinct Pleistocene fauna.

The cranium in the Neanderthal remains was unlike that of any human group alive today. It had prominent eyebrow ridges, a receding forehead and was low in height, all of which gave it an ape-like appearance. In fact it was the most ape-like human cranium yet discovered, Huxley reported in a thorough comparative description of the fossil in *Man's Place in Nature*. Although it was comparatively old and ape-like, Huxley could not quite bring himself to say that the skeleton represented a being intermediate between man and ape. The deciding factor for him was the size of the brain, which had been estimated at over 1000 ml, well within the modern human range and twice that of the largest ape.

Where Huxley hesitated, a geologist by the name of William King stepped in. Neanderthal man was an early forerunner of man, he declared, but had been incapable of moral or rational conceptions on account of the low cranium. This clearly excluded Neanderthal man from the human species, and in 1864 King duly assigned this fossil to a new species, *Homo neanderthalensis*. This was a bold move: no one before had thought of naming a distinct species for our fossil predecessors. Of course this whole line of interpretation met with stiff opposition from other experts, who thought that this individual had lived within historical times. They attributed the skull's peculiarities to idiocy or to a case of rickets, or better still an idiot suffering from rickets. The specimen itself was too incomplete and poorly dated to settle the issue.

But it did not matter. The combined evidence of anatomy and embryology, and of stone implements and fossils, had effectively demolished the picture of man painted by Cuvier half a century before. No longer could it be said that humans were separated in time from the most recent of extinct species. Nor could it be maintained that we were set apart from living species by some unique structure. The new picture suggested unequivocally that man was in nature and not above it. So it came about that Darwin, who had held back earlier, now felt free to enlarge on this new picture. His *Descent of Man* appeared in 1871.

In the midst of all this, there was one important point on which Darwin failed to convince his contemporary naturalists. He could not get them to accept natural selection as the main mechanism by which evolution had come about. This was in spite of the fact that the overall case for evolution was evidently helped by his theory of natural selection. As he had anticipated, people found it easier to accept evolution when they saw he had discovered a cause of evolution that no one had thought of before.

Remains of the original specimen of Neanderthal man on display in the Provincial Museum at Bonn in the late nineteenth century.

Nevertheless, most naturalists doubted that natural selection could do all that Darwin claimed for it.

Once again, Huxley illustrates the general reaction rather well. He thought that Darwin had loaded himself 'with an unnecessary difficulty' in insisting that natural selection always worked slowly and gradually. Evolution might well proceed by sudden jumps every now and again, in Huxley's opinion. And he always maintained that one should be able to breed separate, mutually sterile species by selection in order to give full support to the theory. So Huxley had reservations about the role of natural selection even though he praised Darwin and sprang to his defence. Darwin, in his turn, had reservations about Huxley's understanding of natural selection. After hearing Huxley speak on evolution in 1860, he said that 'as an exposition of the doctrine the lecture seems to me an entire failure', and added: 'He gave no just idea of Natural Selection.'

Part of Darwin's problem was that he had no direct evidence that natural selection changes species in the wild. Strong as his case was, it was indirect. His case was based on the clear effectiveness of selection in domestic breeding and on the argument that the struggle for existence must exert a selective effect in the wild. This was fine as far as it went, but it did not go far enough to convince everybody that evolution took place mainly through natural selection.

There was, however, some new work that came close to providing direct evidence of natural selection in the wild. This was the discovery of mimicry by Henry Bates, who had shared Wallace's adventures in South America. There he had studied insects and paid special attention to the brightly coloured butterflies of the family Heliconidae in the Amazon region. His most interesting finding was that a 'mimetic analogy' existed between some species of Heliconidae and certain species of butterflies from quite different families, such as the Pieridae. Typically, the Pieridae are rather plain white butterflies, but Bates found several species that resembled particular species of Heliconidae in 'external appearance, shape, and colours' (see colour plate 11). The resemblance was so close that it was impossible to tell them apart 'when on the wing in their native forests'.

Bates noticed that each of the mimics among the Pieridae was confined to the same district as the species of Heliconidae that it resembled. Sometimes a widespread species, such as *Leptalis theonoe*, was found to mimic a different species of Heliconidae in different districts. In this case, the mimetic species was subdivided into a number of distinct geographical races, which were often linked by intermediate forms. From facts like these, he decided that mimicry was an adaptation, in which the mimic gained protection from predators by being mistaken for one of the Heliconidae. The latter clearly had some advantage that enabled them to fly slowly through the forest in large numbers, without being molested by insectivorous animals. 'It is probable they are unpalatable', Bates observed, having noticed an unpleasant smell when handling the insects.

By what process could this mimicry be brought about? It could not be caused by local physical conditions since species like *Leptalis theonoe* sometimes mimic two distinct species of Heliconidae in the same locality. It could not have arisen at one jump because different degrees of mimetic accuracy are found. Some active principle must be at work to build up such exact resemblances, each in its own geographical locality. 'This principle can be no other than natural selection,' Bates declared, 'the selecting agents being insectivorous animals, which gradually destroy those sports or varieties that are not sufficiently like [the Heliconid model] to deceive them.' Hence the key to the formation of new adaptations or species is 'numerous small steps of natural variation and selection'. When this paper was published in 1862, Darwin was understandably delighted.

The story was taken a stage further when Darwin came to develop his theory of 'sexual selection' a few years later. He considered that bright colours in animals are of use in courtship and so can be explained by the females constantly choosing the most brightly adorned males. But then he realized that some caterpillars are brightly coloured even though they have no sex life at all. He mentioned the problem to Bates, who said: 'You had better ask Wallace.' Almost by return of post, Wallace came up with the theory of warning coloration, which he had thought out with characteristic insight. The clue that led him to this theory was the realization that these conspicuous caterpillars are cases of protective colouring and not of sexual adornment.

Wallace reasoned that most species of caterpillars have colouring that tends to conceal them from the attention of birds, which eat them in large numbers. The brightly coloured species must therefore carry some protection that allows them to be conspicuous (see colour plate 12A). And since unpalatable butterflies, such as the Heliconidae, are brightly coloured, it seemed likely that the brightly coloured caterpillars were also unpalatable. Further, the conspicuous colouring must serve as a warning to birds so that the insects are left alone rather than being half eaten and then rejected. Wallace therefore predicted that brightly coloured species of caterpillar would prove to be distasteful to birds while all inconspicuous species would be readily eaten.

'Bates was quite right;' Darwin wrote back gratefully, 'you are the man to apply to in a difficulty. I never heard anything more ingenious than your suggestion, and I hope you may be able to prove it true.' Wallace's prediction was shown to be correct when naturalists began to test it using tame birds and lizards. Taken together, then, warning coloration and mimicry did show that the colours of animals are adaptations to their environment, and these adaptive colours made obvious sense in terms of natural selection. The whole topic was worked out in some detail by Wallace and also by Edward Poulton well before the end of the century.

Nevertheless, this was not enough to turn the tide of opinion that was running against natural selection. The main problem was that Darwin's contemporaries could not see that natural selection was capable of producing anything really new. The steady removal of less well-adapted individuals might modify or improve existing adaptations, but how could such a process create a new adaptation?

This difficulty was well expressed in Pictet's review of the *Origin* in 1860. There he wrote that 'normal generation' and natural selection just did not seem adequate to account for the origin of new types of organism. In addition to the force of normal generation and variation, a 'creative force' was needed to produce new types in the course of time. Pictet agreed with Bronn in saying that the nature of this creative force was unknown but it was not supernatural. He explicitly rejected the view that the appearance of each new type in the fossil record was due to divine intervention. In these terms, Pictet agreed that natural selection might produce some changes but it did not qualify as the creative force.

Much the same view was held by Lyell, who never did accept the theory of natural selection, in spite of a long and fascinating exchange of letters with Darwin. In his notes on the species question, Lyell compared the situation with the three attributes of the Hindu deity: Brahma, the creator, Vishnu, the preserver and Siva, the destroyer. 'Natural Selection will be a combination of the two last,' he observed, 'but without the first, or the creative power, we cannot conceive of the others having any function.'

But it was the essence of Darwin's position that natural selection is the creative force of evolution. He tried to persuade Lyell that it only needed plenty of variation and the long-continued action of natural selection to produce new types of organism. Breeders managed to improve their stocks of cattle or pigeons by selecting from copious variation, without the need for any other creative force or power of adaptation. Similarly, Darwin could see no limit to a corresponding process of improvement in evolution. There was no need for the intervention of any additional force. Grant a simple archetypal creature, Darwin wrote with regard to the vertebrates, 'and I believe that natural selection will account for the production of every vertebrate animal'.

In 1871, a compendium of all the main objections that could be raised against natural selection was published by St George Mivart in *The Genesis of Species*. As others had done before him, Mivart (1827—1900) accepted the case for evolution but doubted Darwin's mechanism. He agreed that natural selection acts in the wild, and is capable of explaining routine examples of evolutionary change. But then he added that, to account for anything really new, it must be 'supplemented by the action of some other natural law or laws as yet undiscovered'.

The most telling point that Mivart brought forward was 'that "Natural Selection" is incompetent to account for the incipient stages of useful structures'. Examples that he cited include the long neck of the giraffe, the baleen plates of whales and the detailed resemblance of some insects to other objects. Surely, complex adaptations like these could not have got started through the preservation and accumulation of small successive variations. How could natural selection possibly produce the first stages of such structures, before they were sufficiently developed to serve their present purpose?

Darwin saw the force of this objection and devoted part of a chapter to answering it in the sixth edition of the *Origin* in 1872. For many of Mivart's examples, he could show that graded intermediates between an incipient stage and a highly developed stage are found in different living species. Since functional intermediate stages actually exist, the evolution of these complex structures could have been produced by natural selection from the start. This would sometimes have been accompanied by a change of function, since structures that originally evolved for one function may then be converted for another function.

The case of insects that gain protection by closely resembling some other object was easily explained. Darwin argued that all it needed to get this started was 'some rude and accidental resemblance to an object commonly found' in the insect's environment, such as a leaf or twig. Then any variations that improved the resemblance at all, and so aided the insect's escape, would tend to be preserved while any variations that spoiled the resemblance would be eliminated. Natural selection would continue to improve the resemblance for 'as long as the insect continued to vary, and as long as a more and more perfect resemblance led to its escape from sharp-sighted enemies'.

A comparison of the difficulties raised in the *Genesis of Species* with the explanations given in the sixth edition of the *Origin* shows Darwin's mastery of the subject. There can be little doubt that he got the better of the argument, but this is not to say that he convinced people that natural selection is the creative force in evolution. In the absence

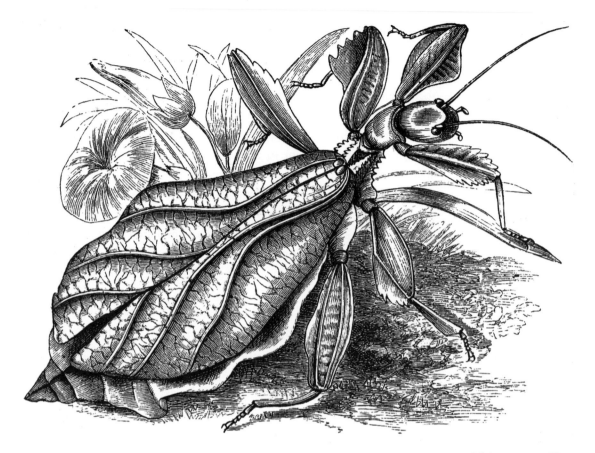

of more direct evidence in favour of natural selection, Mivart's view that it must be supplemented by some other mechanism held a wide appeal.

Mivart thought this supplementary mechanism must involve a kind of 'internal innate force' to push variation in some definite direction. And this directed variation must sometimes be large enough to get from one species to another in a single jump. With this suggestion, he set a pattern for alternatives to natural selection that was to last well into the twentieth century. Either directed variation or big jumps or both together were increasingly seen as the main mechanism of evolution, with natural selection relegated to a subordinate role.

The crucial issue here was the nature of variation. Darwin recognized that variation must have certain properties for natural selection to play a creative role in evolution. The main requirement was that variation should be plentiful, since a mechanism of selection needs lots of raw material from which to choose, if it is to be effective. From his experience with barnacles and pigeons, Darwin knew that individuals of a species do

A stick insect resembling a leaf, from Mivart's Genesis of Species. Mivart doubted that natural selection could account for the incipient stages in the evolution of such resemblances.

Plate 13 Twenty-five species of snail in the genus Achatinella *from Oahu in the Hawaiian Islands, illustrated by John Gulick in 1905. The small letter beneath each shell indicates the valley from which it was collected, as shown in the map of Oahu on page 193. Notice how each shell tends to resemble the ones immediately above and below it, i.e. those from neighbouring localities.*

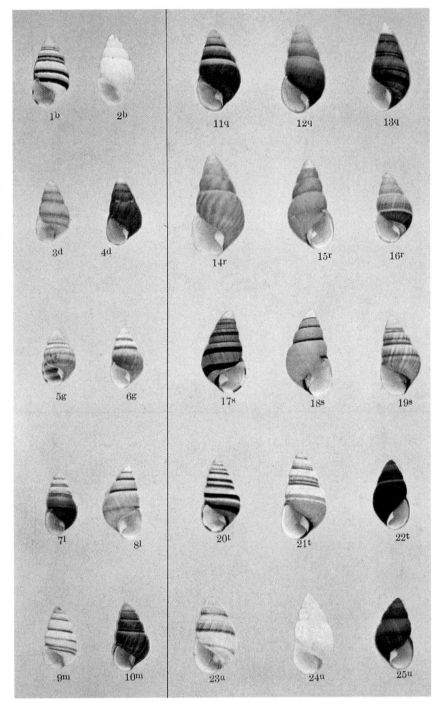

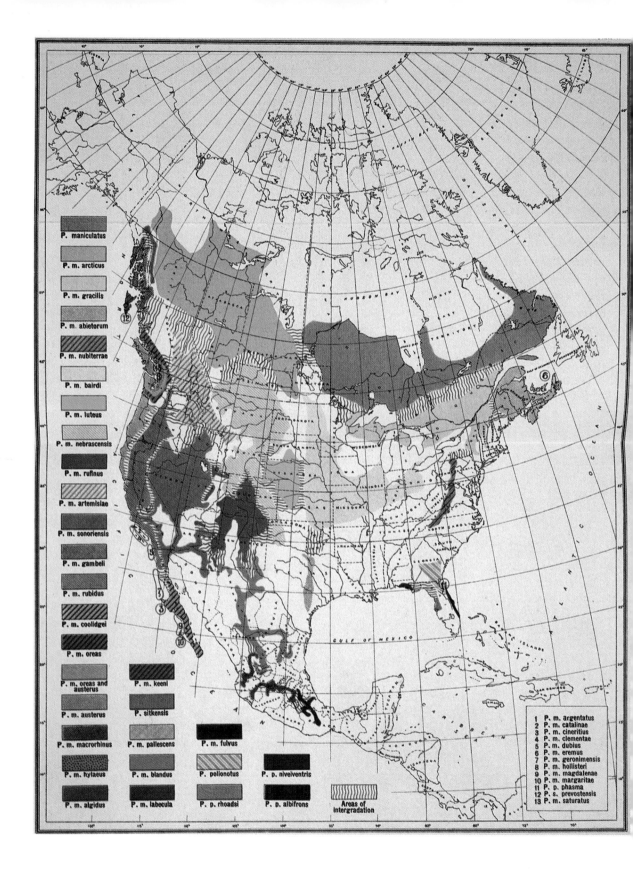

P. maniculatus

P. m. arcticus

P. m. gracilis

P. m. abietorum

P. m. nubiterrae

P. m. bairdi

P. m. luteus

P. m. nebrascensis

P. m. rufinus

P. m. artemisiae

P. m. sonoriensis

P. m. gambeli

P. m. rubidus

P. m. coolidgei

P. m. oreas

P. m. oreas and austerus

P. m. keeni

P. m. austerus

P. sitkensis

P. m. macrorhinus

P. m. pallescens

P. m. fulvus

P. m. hylaeus

P. m. blandus

P. polionotus

P. p. niveiventris

P. m. algidus

P. m. labecula

P. p. rhoadsi

P. p. albifrons

Areas of intergradation

1   P. m. argentatus
2   P. m. catalinae
3   P. m. cineritius
4   P. m. clementae
5   P. m. dubius
6   P. m. eremus
7   P. m. geronimensis
8   P. m. hollisteri
9   P. m. magdalenae
10 P. m. margaritae
11 P. p. phasma
12 P. s. prevostensis
13 P. m. saturatus

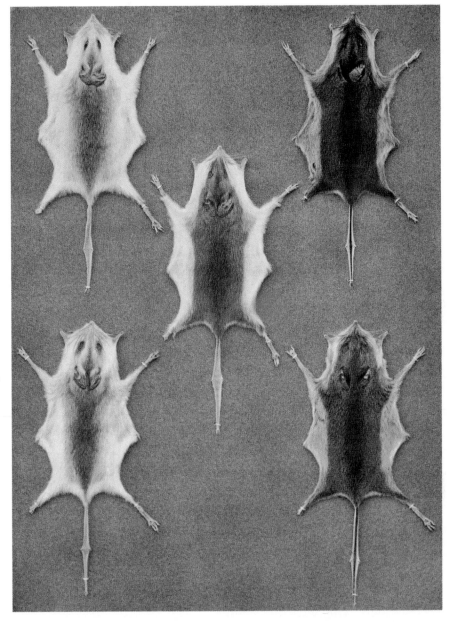

Plate 14 A map published in 1909 showing the distribution of subspecies in the mouse, Peromyscus maniculatus, *and a few closely related species. Each subspecies is confined to its own region but there some areas of intergradation between neighbouring subspecies as indicated on the map. The numbers refer to subspecies that are confined to small, isolated localities, mostly islands.*

Plate 15 The genetics of coat colour in the mouse, Peromyscus polionotus. *When Sumner crossed the pale race,* leucocephalus *(TOP LEFT) with the dark race,* polionotus *(TOP RIGHT), the F_1 generation was intermediate in appearance. But the extremes of variation in F_2 (BOTTOM LEFT and RIGHT) closely resembled the grandparental races, so providing evidence for Mendelian segregation.*

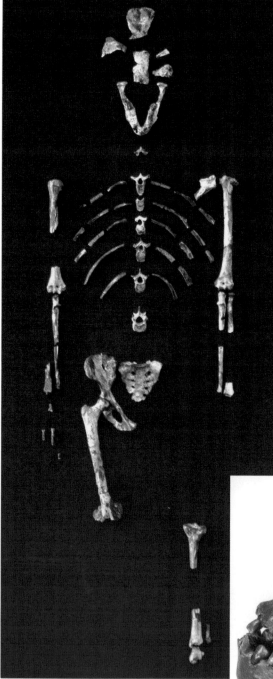

Plate 16A The remarkably complete skeleton of 'Lucy', a fossil hominid discovered in the 1970s. Formally assigned to the species Australopithecus afarensis, along with other fossils from the same locality, this find was an important milestone in research on human origins.

Plate 16B Skull of the fossil hominid known informally as Turkana Boy. The skeleton of Turkana Boy resembles that of modern humans from the neck down but the skull differs in several features such as its brow ridges, receding forehead and somewhat protruding jaws.

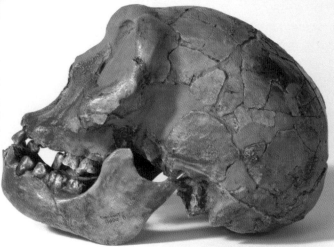

differ from each other in many points of detail. He emphasized that these variations are small in extent. Variations that were large enough to generate a new species in a single jump would render natural selection unnecessary.

Darwin also maintained that variation took place equally in all directions, with no preference for a useful direction. Although he wrote of natural selection acting on 'numerous, successive slight favourable variations', it was an essential part of his theory that variations are neither favourable nor unfavourable in themselves. They merely happen, and their favourable or unfavourable qualities only show up in the competitive struggle for existence. So Darwin's view was that variations are small, plentiful and undirected. Consequently, the only way to evolve a new structure or species is by the slow cumulative selection, generation after generation, of variation that happens to go in that direction. Most of Darwin's contemporaries were reluctant to grant this creative role to natural selection because they did not accept this view of variation.

Yet on this topic, Darwin himself never entirely broke away from the old view. He kept harking back to the idea that variation is somehow unnatural, and that the external conditions of life influence both the amount and direction of variability within a species. Right up to the sixth edition, the *Origin* was sprinkled with qualifying phrases such as 'if variations useful to any organic being do occur ...'. But Wallace, who had no such hesitations, wrote to say that 'it would be better to do away with all such qualifying expressions'. Instead Darwin should maintain that 'variations of every kind are always occurring in every part of every species, and therefore that favourable variations are always ready when wanted'. This, urged Wallace, is 'the grand fact that renders modification and adaptation to conditions almost always possible'.

In this same letter, which was written in 1866, Wallace offered Darwin another piece of advice. He pointed out that those new to the subject often made the mistake of thinking that the term natural selection implied some sort of conscious choice by a personified 'Nature'. And he suggested that this misconception could be avoided by adopting Herbert Spencer's term, 'survival of the fittest', as a synonym for natural selection. Darwin followed this advice in later editions of the *Origin*, and so the phrase 'survival of the fittest' became part of our language, not always without confusion. However, as far as variation was concerned, Darwin did not noticeably alter his wording along the lines suggested in Wallace's letter.

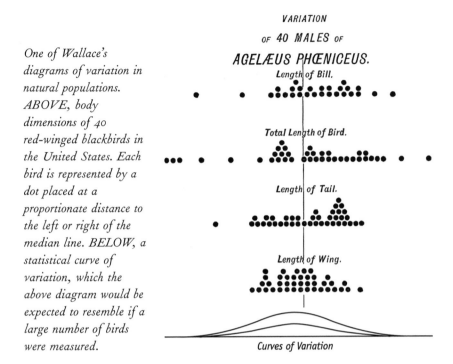

One of Wallace's diagrams of variation in natural populations. ABOVE, body dimensions of 40 red-winged blackbirds in the United States. Each bird is represented by a dot placed at a proportionate distance to the left or right of the median line. BELOW, a statistical curve of variation, which the above diagram would be expected to resemble if a large number of birds were measured.

Wallace saw the shape of ideas to come more clearly. He continued to stress the amount of variability found in wild populations, in opposition to those who insisted that variation is comparatively rare in nature. A good case was made out in his popular book on evolution by natural selection, which he entitled simply *Darwinism*, in 1889. With the help of special diagrams, he gave a quantitative measure of variation in natural populations of reptiles, birds and mammals. This method showed that variation was always present and that it was not so very small in extent. For any given part of the body, individuals commonly varied by as much as 20% of the mean value.

Important though this evidence was, it was not on its own enough to settle the question of variation, for it did not take account of how much of the variation was inherited. An understanding of the causes of variation and inheritance was therefore needed to complete the picture. For forty years or so after the *Origin* first appeared, the lack of such an understanding was the biggest single gap in the theory of evolution. In the *Origin* itself, Darwin had been content just to call on 'the strong principle of inheritance', and so sidestepped all questions about its mechanism. But he knew all along that a knowledge of the mechanism of inheritance was needed to complete the theory of natural selection.

So strongly did Darwin feel the need to complete the system of causes that he came up with a 'provisional hypothesis' to explain how inheritance works. This theory, called 'pangenesis', was published in the *Variation of Animals and Plants under Domestication* in 1868. The basic idea was that tiny particles called 'gemmules' or 'pangenes' are given off by every part of the adult body. These gemmules circulate round to the reproductive organs, where they are incorporated into the germ cells — the eggs and sperm in animals. These cells, from which the offspring develop, thus carry hereditary particles produced by all parts of the parents' bodies. Consequently, the characteristics of the offspring will reflect the condition of the parents at the time of conception.

This theory tried to explain inheritance in terms of the old idea that inheritance involves something made by the parents that is passed on to the offspring. Such a theory would certainly explain how acquired characteristics could be inherited, which was an idea that most naturalists still took for granted. Changes that took place in some part of the body during the parents' lifetime would affect the gemmules given off, and so would be passed on to the next generation. Darwin accepted this but thought that most variation would not be directed in this way; usually variation would result from gemmules being altered haphazardly by external conditions that impinge on the reproductive organs.

Pangenesis was a well thought out theory of inheritance but it had only a short life. It stimulated research on the link between evolution and heredity, and soon became obsolete as a result. One person who went out of his way to test the theory was Darwin's cousin, Francis Galton (1822–1911). He made blood transfusions between rabbits that belonged to pure breeds with different coat colours. He then bred from the transfused rabbits to see if there was any trace of the other breed's coloration in the offspring, as there should have been if gemmules had been circulating in the blood. But the offspring showed no departure from the parental coat colours, and this confirmed doubts that Galton already had about pangenesis.

From the constancy of human races over a wide range of geographical conditions, Galton was convinced that hereditary traits are transmitted unchanged through many generations. He felt sure that an individual's inheritance is derived, not just from the parents, but from all the ancestral generations. A consequence of this 'law of ancestral heredity' was that the hereditary material in a fertilized egg must be passed on unchanged from one generation to the next. It could not be manufactured by the parents under the influence of local conditions in the way that Darwin and earlier naturalists had imagined. This meant that traits acquired by the parents during their lifetime could not be inherited to any significant extent.

Galton began to study heredity in terms of a whole population, and to do this he employed new statistical methods, including the techniques of correlation and regression. By thinking in terms of a population, it became possible to see that heredity and variation are simply two sides of the same coin. Variation is not a force that interferes with heredity, but is simply the preservation within a population of a range of traits, all inherited from distant ancestors. Offspring may differ from their parents simply because they show a trait transmitted from a more distant ancestor, not because local conditions have interfered with the accurate copying of parental traits.

As to the basis of heredity, Galton shared Darwin's view that it must be effected through material particles carried in the germ cells during reproduction. But he did not try to find out more about exactly how heredity transmission is effected. He kept to the study of heredity in large populations, which in itself was of benefit to a whole generation of biologists. Insight into the basis of heredity came instead from research on the microscopic study of cells, a field with which Darwin and Galton were unacquainted. This insight was to become a major step forward for biology as a whole, but its initial effect on the study of evolution was to generate controversy and confusion.

CHAPTER SEVEN

A RICH INHERITANCE

Sometimes in life we are offered the solution to a vexing question but at first we fail to recognize it as such. If anything, it seems to make the problem worse. This is exactly what happened with the study of inheritance in relation to evolution, as the nineteenth century passed into the twentieth. The theory of natural selection had been proposed by Darwin and Wallace to explain the process of evolutionary change. It met with only partial agreement from other naturalists, and a number of alternatives were put forward. What was needed to clarify the problem was a fully worked out theory of heredity. But when this was first achieved, rather than making the process of evolution clear, it caused more confusion.

While Galton was using statistical methods to study heredity in populations, the possible basis of inheritance was being revealed by the microscopic study of cells. Robert Hooke had been the first to observe cells under the microscope some two centuries earlier, and had given them their name in his *Micrographia*. However, it was not until early in the nineteenth century that the advent of better microscopes brought the cell quite literally into focus. Then the microscopical observations of Matthias Schleiden and Theodor Schwann enabled them, in 1838–9, to propose the theory that all plant and animal tissues are composed of cells. Further work showed that each cell consists of three basic components: an outer membrane, the enclosed cell contents and a central kernel or 'nucleus'.

The process by which a cell divides was also observed, and was incorporated in the cell theory by Rudolph Virchow. In 1858 he argued that cells arise only from the division of pre-existing cells, and cannot crystallize out of a structureless fluid as Schwann had supposed. By 1875, other microscopists could add to this by showing that each nucleus is formed out of a pre-existing nucleus during the process of cell division. Following the demonstration that the egg and sperm are each single cells, they showed that it is the nuclei of these cells that unite at fertilization. Hence it must be the nucleus that carries the basis of heredity from one generation to the next.

At this point, the matter was taken up with great skill by August Weismann (1834–1914), who had made a name for himself in microscopical research. When he found that he could no longer work at the microscope because of failing eyesight, Weismann turned his attention to the problems of heredity and evolution. Beginning around 1880, he published a closely reasoned series of discussions on heredity, culminating in his book on *The Germ Plasm* in 1892. He realized that Darwin's theory of pangenesis could not be reconciled with what was known about cell division. The hereditary material or 'germ plasm' was evidently carried in the nucleus of the 'germ cells', the egg and sperm (also called gametes). Hence the germ plasm must be derived from the division of pre-existing cells and could not be an extract or product of the whole body.

In fact, Weismann decided, the germ plasm must be a particular chemical substance that was part of the cell nucleus. As he stated clearly, his theory was 'founded upon the idea that heredity is brought about by the transmission from one generation to another of a substance with a definite chemical and, above all, molecular constitution'. It was this molecular structure of the germ plasm that enabled a germ cell 'to become, under appropriate conditions, a new individual of the same species'. He reasoned that the germ plasm must be divided into many units, the 'determinants', each of which controls the development of a certain part of the body.

With this theory, new observations on the nucleus of a dividing cell made sense at once. The nucleus was shown to contain material, the 'chromatin', that coloured strongly with the new synthetic dyes. When a cell was about to divide, the chromatin became concentrated in a number of rod-shaped 'chromosomes' (Greek for 'coloured bodies'). As the cell divided, the chromosomes were duplicated by splitting longitudinally; one set of chromosomes then passed to each daughter cell and there formed part of a new nucleus. Weismann pointed out that this complex mechanism, which was later called mitosis, exists 'for the sole purpose of dividing the chromatin' in a fixed and regular manner. This showed that the chromatin is the most important part of the nucleus and, therefore, 'the chromatin must be the hereditary substance'.

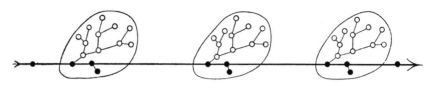

Diagram illustrating the continuity of the germ plasm from one generation to the next. Successive individuals (large ovals) are formed from body cells (white circles) but the germ cells (black circles) form a separate lineage, not derived from the rest of the body.

If this were so, Weismann argued, the germ cells must be formed by a cell division that took place without duplicating the chromosomes. This would produce egg and sperm cells having only half the normal number of chromosomes. The fusion of egg and sperm at fertilization would then restore the chromosome number, ready for the subsequent development of the embryo by normal mitotic division. This idea was confirmed in the late 1880s, when the special kind of division by which the germ cells are formed was observed under the microscope. Now known as reduction division or 'meiosis', it involves an initial duplication of the chromosomes followed by two successive divisions without any duplication. Hence these new findings all supported the theory that heredity is based on a chemical substance carried in the nucleus.

As he sought to develop his theory of the germ plasm, Weismann was all the while thinking about its consequences for evolution. From the start, he was convinced that natural selection must be the main cause of evolutionary change, and this led him to study warning coloration and mimicry in insects. His views were reinforced by microscopical observations that he had made on some small marine animals, the Hydromedusae. There he found that the embryonic cells which are going to give rise to germ cells are set aside early in development. They do not contribute to the growth of the rest of the body, but form quite a separate lineage of cell divisions. This appeared to rule out the possibility that traits acquired by the body could be passed on to the next generation by the germ cells.

The transmission of hereditary traits was therefore unrelated to the growth of the organism, as Galton had perceived from a different perspective. And like Galton, Weismann carried out an experiment to test this conclusion. He cut off the tails of mice, generation after generation, and showed that there was no tendency for the mutilations to be inherited. Hitherto, it had been widely accepted that mutilations could be inherited, and this was taken as proof of the inheritance of acquired characters. Other studies bore out Weismann's results, and in his last book *The Evolution Theory* in 1904 he could say confidently: 'we may now regard this kind of "proof" as disposed of'.

Another reason for rejecting the inheritance of acquired characters was that Weismann could not imagine any process by which it could happen. To be inherited, the enlargement of a muscle through use must somehow be reflected in the germ plasm, which is not composed of muscle. The variation in the body would have to be turned into a different sort of variation in the germ plasm, 'which is very like supposing that an English telegram to China is there received in the Chinese language'. As this analogy shows, Weismann saw that what matters in heredity is not a flow of matter or energy but of information. And there did not appear to be any mechanism by which information from the body could be translated into information in the germ plasm.

If bodily modifications acquired during growth could not be inherited, then variation must be undirected. Normal variation, Weismann explained, is due to mixing of the determinants in the germ plasm as they are shuffled by sexual reproduction. New variation can only arise by alterations in the molecular structure of the germ plasm caused by accidental errors in duplication. But the body has no control over such accidents, and the new variations occur at random. Natural selection is therefore the only factor that can give an adaptive direction to hereditary changes during the course of evolution. From this position, Weismann claimed that natural selection was 'all sufficient' as a cause of evolutionary change.

However, the dogmatic way in which he advocated this view did not help to get it generally accepted. His approach tended rather to stiffen the resolve of those who had doubts about the role of natural selection in evolution. Prominent among these was the philosopher, Herbert Spencer, who wrote an essay opposing Weismann's views in 1893. Spencer argued that Weismann's evidence was inconclusive, and insisted that variations could be directed along favourable lines through the inheritance of acquired characters. So there grew up two opposed camps: the strict Darwinians, who argued for the all-sufficiency of natural selection, and the Lamarckians, who promoted the inheritance of acquired characters and other mechanisms of evolution.

A good example of a Lamarckian was Edward Cope, who based his interpretation of vertebrate evolution on the discovery and skilled interpretation of many new fossils. In studying the succession of these fossils over time, Cope followed a natural inclination to simplify the evolutionary trends by treating them as straight lines. One example here was the regular transformation of horns, teeth and feet in the hoofed mammals as they evolved from the generalized condition seen in an early form like *Phenacodus*.

Such a pattern of evolution, advancing step-by-step along a number of straight lines, could not be explained by natural selection in Cope's view. Selection of random variations should result in an irregular pattern, so these regular trends must be due to some other mechanism. He came to the conclusion, possibly through the influence of Spencer, that the inheritance of acquired characters would result in linear evolution. Since his fossil evidence seemed to require this interpretation, Cope took little notice of the sort of evidence about inheritance marshalled by Weismann.

There were others, however, who found the new climate created by Galton and Weismann conducive to further study. If the mechanisms of fertilization responsible for inheritance were so simple and precise, should this not be reflected in regular patterns of inheritance? And if the germ plasm should happen to be altered, would this not show up as a discontinuity in the pattern of inheritance? During the 1890s a number of people began to tackle the problems of heredity from this new point of view, leaving the problems of growth and development on one side.

One such person was the British zoologist, William Bateson (1861—1926), who had begun his career in comparative anatomy. He had hoped to throw light on the origin of vertebrates by studying the simple, worm-like creature, *Balanoglossus*. Eventually he came to feel that this was a futile exercise because the available facts were open to more than one interpretation, with no way of deciding between them. Seeking a more fruitful approach to the study of evolution, he turned to look at the raw material for evolution, that is at variations. The outcome of this line of work was the publication of his *Materials for the Study of Variation* in 1894.

Here he argued that the best way of studying evolution was to study the facts of variation. 'Variation,' he wrote, '. . . is the essential phenomenon of Evolution. Variation, in fact, *is* Evolution.' Bateson adopted the view that there are two kinds of variation, which he called 'continuous' and 'discontinuous'. The former term meant slight individual differences that make up the curve of variation for a species, as in the cases stressed by Wallace. The latter term meant striking differences that are not part of a range of variation. In the *Origin of Species*, Darwin had relied on continuous variation as the basis for gradual evolution through natural selection, but Bateson turned to discontinuous variation as the basis for evolution by sudden jumps. As he put it, 'Species are discontinuous; may not the Variation by which Species are produced be discontinuous too?'

Bateson's evidence for this view was drawn from the large amount of material he had gathered on natural species. He cited lots of cases where two or more distinct forms were found in the same species, without any sign of intermediates. Such cases were especially clear where they involved differences in the number of similar parts, as in the number of petals in a flower or of segments in a body. For intermediates were simply not possible with these numerical differences. Moreover, variations of this kind were about the same magnitude as the differences that distinguish one species from another. This suggested 'that the Discontinuity of Species results from the Discontinuity of Variation'; in other words, that a new species arises at one jump from a new variation.

This point of view was not entirely new, of course, but Bateson took it further than earlier naturalists. As he developed this view, he realized that discontinuous variations would each have to be inherited as a discrete unit if they were to play the role he assigned to them. This kindled his interest in the hereditary transmission of variations. So in the 1890s he began a series of experiments in which he cross-bred individuals of a species showing discontinuous variations to see how these were inherited.

Another person who took up breeding experiments at this time was the Dutch botanist, Hugo de Vries (1848–1935). He came to the study of heredity from a background in chemistry, and from the first his main interest was in the transmission of hereditary traits and evolution. In 1889 he published *Intracellular Pangenesis*, a book that modified Darwin's views on heredity in the light of the new understanding of cells. In particular, it developed the idea of a hereditary unit underlying each trait or character of a species. The attributes of a species were not indivisible, said de Vries, but could be separated into a large number of unit characters. Each of these must be due to a single hereditary factor, which was more or less independent of the others. These factors 'are the units which the science of heredity has to investigate', just as physics and chemistry are based on the study of atoms and molecules.

De Vries called these hereditary units pangenes, after Darwin's original theory, and argued that each one can vary and be inherited independently

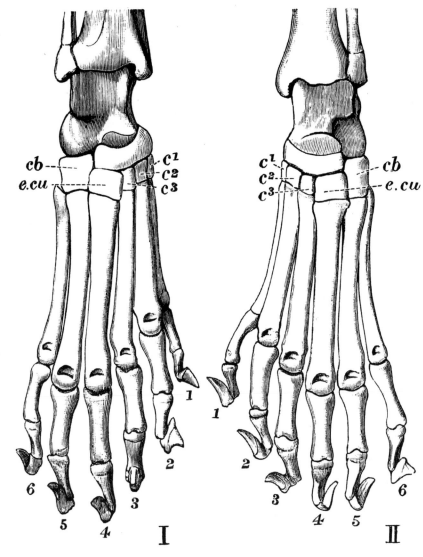

Hind feet of a cat with six digits on each foot, cited as an example of discontinuous variation by Bateson. Normally, cats have only four well-developed digits on each hind foot, rather the usual five, due to the innermost digit (hallux) being either absent or a mere rudiment.

of the others. Here he differed from Weismann, who could never quite countenance independent units within the germ plasm. If the pangenes were independent units, then it should be possible to follow each one through successive generations by means of breeding experiments. So in 1892 de Vries began to breed a number of common plant species, crossing individuals that differed in some conspicuous character. He soon found, as Bateson was doing in England, that this procedure gave rise to a consistent pattern of inheritance.

By the autumn of 1899, de Vries had obtained a similar pattern in some thirty different species and varieties. He was convinced that he had found a general law for the transmission of hereditary traits, and began to prepare his work for publication. As he looked back through the literature, he discovered to his surprise that he had been anticipated. More than thirty years before, in 1866, a monk by the name of Gregor Mendel had reached the same conclusion, but his work had been overlooked.

There can be few men who have made a posthumous contribution to science but Mendel (1822–84) was one. He had been educated at Vienna University, where he studied subjects ranging from physics to botany and became acquainted with the methods of experimental research. With this background, it seems that Mendel soon developed an interest in the long-standing problems of plant hybridization and heredity. He decided to tackle the matter experimentally by breeding plant hybrids and analysing the inheritance of their distinctive traits in a long-term study.

With this in mind, he began breeding experiments using the garden pea, which was chosen for its suitability. To start with, he identified seven characters that exist as a clear-cut pair of alternatives, such as tall or short stems, round or wrinkled seeds. These were examples of what Bateson would later call discontinuous variation, and Mendel chose them so that he could readily follow the transmission of the characters through several generations.

Mendel began with plants that bred true for each of the seven traits; he then crossed two plants differing in one pair of alternative traits and examined all the offspring. He found that in the first hybrid generation (F_1) all the plants resembled one of the parents and not the other. For instance, when tall and short plants were crossed all the F_1 plants were tall. Mendel described the character that appeared in F_1 as 'dominating' and the alternative that had apparently disappeared as 'recessive'. These hybrids were then self-fertilized, and the seeds grown to maturity, to produce a second hybrid generation (F_2).

In this second generation, Mendel found that the recessive traits reappeared, and he showed that the proportion of dominant traits to recessive ones was $3:1$. He could be sure about this ratio because he worked with such large numbers and analysed all his results mathematically. In the case of seed shape, for example, 7324 seeds collected from 253 plants gave 5474 round and 1850 wrinkled (a ratio of $2.96:1$). He did not stop here, but produced a third generation (F_3) by self-fertilizing a large

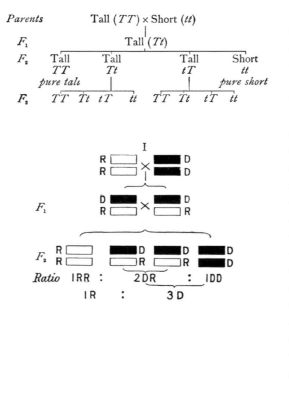

Parents		Tall (TT) × Short (tt)		
F_1		Tall (Tt)		
F_2	Tall TT *pure tall*	Tall Tt	Tall tT	Short tt *pure short*
F_3	TT Tt tT tt		TT Tt tT tt	

Two diagrams explaining the Mendelian pattern of inheritance, from Mendel's Principles of Heredity by William Bateson. ABOVE, the results of crossing tall and short peas represented in the conventional notation. BELOW, the numerical consequences of segregation made clear by representing each zygote as a pair of draughts, with the black draughts being dominant and the white draughts recessive.

number of plants from F_2. All the plants raised from wrinkled seeds bred true for this character, as did one-third of the plants raised from round seeds. The other two-thirds of plants raised from the round seeds gave round and wrinkled seeds in the proportion of 3 : 1.

Mendel explained this regular pattern of inheritance by supposing that the F_1 generation contained hereditary 'elements' for both of the alternative traits. The recessive element was masked but was none the less available for transmission to the next generation. Then there was an independent assortment of the two types of element in the formation of the F_1 germ cells, so that four possible combinations could occur in plants of the F_2 generation. Three out of the four combinations would carry the dominant element, and so give rise to the 3 : 1 ratio. Writing in 1866, Mendel was understandably vague about his hereditary 'elements', but by 1900 they could be interpreted in terms of the cell biology outlined by Weismann.

The full significance of Mendel's pioneering work for the science of heredity was abundantly clear to Bateson. He saw that the 3 : 1 ratios could only be explained by assuming that each trait was transmitted from one generation to the next by a single hereditary unit in the germ plasm. This unit must exist in two alternative forms, corresponding to the pair of alternative traits. The fertilized egg or 'zygote' formed by the original cross will then have received one hereditary unit for, say, tallness (T) from one parent and one for shortness (t) from the other. The F_1 plant that grows from this zygote will carry both units (Tt), which must separate when its germ cells are formed. Each germ cell then carries only one unit, either T or t, and the F_2 zygotes will contain two units in four possible combinations of T and t (TT, Tt, tT and tt). The ratios of tall and short plants in F_3 can be accounted for in the same way.

Bateson applied the term 'segregation' to this separation of hereditary units in the course of germ cell formation, and he saw that this was the crucial point in Mendel's work. It was segregation that gave rise to definite ratios in the hereditary transmission of traits and to the discontinuity of so much variation. That each trait is represented in a zygote by just two units, one derived from each parent, and that these remain separate, was the new idea that would revolutionize the study of heredity. Bateson immediately seized the opportunity that this discovery provided for bringing order to a field of study where previously there had been so much confusion.

One thing he did was to introduce several key terms, including the word 'genetics' for this new science of heredity. To the paired hereditary units that segregate from each other he gave the name 'allelomorph', later shortened to 'allele'. A zygote formed from two germ cells each bearing the same allele he termed a 'homozygote' (such as TT or tt), and one with two different alleles a 'heterozygote' (Tt). Bateson promoted the cause of genetics with great energy, and people started breeding experiments on a wide range of plant and animal species. A Mendelian pattern of inheritance was found to apply in all of them, and it could even be shown to extend to complex cases, which at first sight did not follow any simple pattern. An outline of this new work was enough to fill a substantial book, *Mendel's Principles of Heredity*, which Bateson published in 1909 (see colour plate 12B).

Meanwhile, de Vries was less enthusiastic about this new field of study. He was more interested in how new species are formed, and the Mendelian pattern of inheritance did not seem to throw any light on this question. He agreed with Bateson that there are two kinds of variation and that no amount of continuous variation could turn one species into another, even with intense selection. A new species must arise from the sudden appearance of some new, discontinuous variation. Over the years, therefore, de Vries had kept a look-out for species that threw up this kind of variation.

In 1886 he had found the evening primrose, *Oenothera lamarckiana*, growing wild in a meadow, and two of these plants differed markedly from the normal type. When taken into cultivation, these two plants bred true, and other new types appeared among the offspring of the normal plants. It seemed to him that the species was disintegrating into a number of forms so different from the parent form as to justify their description as new species. De Vries introduced the term 'mutation' for the process by which these new forms had originated. His theory that new species arise at one jump through such mutations was published in two volumes, entitled *The Mutation Theory*, between 1901 and 1903.

Later on, it was shown that the evening primrose is a highly unusual hybrid species and that de Vries' mutations were not at all what he thought them to be. He himself had examined more than 100 other species that did not behave like the primrose, but he set these aside as being 'in an

A flower of Dahlia variabilis, *in which each floret has the unusual form of a long, broad tube. This variant individual appeared in a crop raised from seed by de Vries, and its distinctive features were inherited by the next generation. de Vries thought this illustrated the way in which new species arise.*

immutable period'. Although his theory was unfortunately built on a single exceptional species, it did focus attention on the question of the origin of new hereditary units. Given that hereditary transmission was effected by numerous separate units, he argued that each new unit added to those already present must form a distinct step. 'The occurrence of a new unit signifies a mutation', he wrote, and he drew a clear distinction between these mutations, which were heritable, and mere 'fluctuations' caused by the environment, which were not.

It was soon perceived that this idea could be welded to the Mendelian view of inheritance. Once they had arisen, new mutations would be inherited in the normal way, and there was no reason why they should not persist provided they were not directly harmful. Their eventual fate would be determined by natural selection weeding out the least well-adapted of these new forms. But since it was envisaged that completely new species could arise by mutation, it followed that natural selection played only a minor role in evolution. It was mutation that brought about all significant change in evolution by giving rise to new genetic discontinuities.

On this view, adaptation and natural selection played no part at all in the origin of new species. In the light of the new knowledge, Bateson declared, the creative role for natural selection advocated by Wallace, Weismann and others 'must be finally abandoned'. The concept of evolution proceeding through the gradual accumulation of small changes in a large number of individuals 'is one that the study of genetics shows immediately to be false'. In the United States, de Vries' theory was promptly taken up by Thomas Morgan, then at the beginning of his career in genetics. His book on *Evolution and Adaptation*, published in 1903, argued that there was no significant role for selection: the course of evolution was determined primarily by the kinds of mutation that occurred.

It is not surprising, then, that these views came to be seen as a complete alternative to the theory of natural selection. Many naturalists who studied species and their varieties in the field could only smile at the idea of new species originating by a sudden jump. They saw evidence for adaptation and gradual variation everywhere and were not about to be put off. Writing to his friend Edward Poulton in 1904, Wallace called mutation 'a miserable abortion of a theory'. And in introducing a book of essays in 1908, Poulton himself poured scorn on the idea of mutation without selection. So the Darwinian camp found itself engaged in a three-cornered fight, against the Lamarckians on the one hand and the Mendelians on the other.

The main weakness of the Mendelian position, as the field natura-lists saw it, was that it gave no explanation of continuous variation. The new techniques worked well with discontinuous variation but not appar-ently with continuous variation, which naturalists thought was most important in evolution. This led some naturalists to doubt whether the Mendelian rules of inheritance were generally applicable. In a book on *Evolution and Animal Life* in 1907, two American naturalists, David Jordan and Vernon Kellog, wrote that Mendel's laws 'by no means apply to all ... cases and categories of inheritance'. Cases of continuous variation, they felt, needed to be explained by 'something besides the Mendelian principle'. The way out of this difficulty was soon provided by genetical research that did throw light on the nature of continuous variation.

The Danish biologist, Wilhelm Johannsen (1857–1927), worked with the garden bean, *Phaseolus vulgaris*. The plants of this species are normally self-fertilized and so tend to be largely homozygous. Johannsen examined all the descendants of a single self-fertilized plant, in nineteen separate cases, with careful statistical analysis. Within each of these 'pure lines', the beans varied continuously in size over a certain range, and he raised plants from both the largest and the smallest beans. He found that the vari-ation among the offspring was virtually identical, regardless of the size of the parent bean, and that this range of variation remained constant through successive generations. This showed that the genetic constitutions of all the beans in a pure line must be the same. The variations in bean size were evidently fluctuations due to differences in their environment, and were not inherited.

In the light of these results, Johannsen introduced a number of new terms. In 1909 he coined the word 'gene', shortened from the pangene of de Vries, to denote the unit of heredity. From this he derived the word 'genotype' for the sum total of genes in a zygote. And as a counterpart to the term genotype, he introduced the word 'phenotype' for the bodily characteristics of an organism. This terminology made it easy to express the important distinction between an organism's genetic endowment and its physical appearance. This distinction had in a way been implicit in the theories of Weismann and de Vries, but the new terminology made it quite explicit. So the new terms were a help in understanding the process of evolution and the role of continuous variation in particular.

Thus the largest and smallest beans within a pure line could be said to have the same genotype, but to have developed different phenotypes due to differences in their environment. Johannsen showed that when all his beans were mixed together, the total population seemed to have only a single phenotype, with a smooth curve of variation. But in fact, different pure lines showed different mean values of bean size, which bred true and so must reflect different genotypes. Hence the total population was made up of several different genotypes, but the distinctions between them were smoothed out by the variation in their phenotypes due to the environment. The result was continuous variation in bean size, which masked the underlying genetic differences.

Such a view was reinforced in 1909 by the work of a Swedish biologist, Nilsson-Ehle, on wheat. While studying the inheritance of colour in the grain, he found that the red colour varied in intensity but red was always dominant to white. Families formed by crossing reds and whites fell into three groups, in which F_1 reds gave reds and whites in F_2 in proportions of either 3 : 1, 15 : 1 or 63 : 1. Nilsson-Ehle showed that these results could be explained if there were three separate genes, each of which could produce the red effect. Given three genes segregating independently, say A, B and C, then pure whites would occur only in plants homozygous for all three recessives (aabbcc). Just one dominant would be enough to turn the grain red. But the depth of colour and the proportion of reds in F_2 would increase with number of dominants present.

It was clear from these results that continuous variation in size or colour could be explained in Mendelian terms where a character was produced by multiple genes. Their interaction would result in graded phenotypes, each a little different from the next. During growth, these differences would be smoothed out by environmental influences to give a continuous curve of variation. There was therefore no essential distinction between continuous and discontinuous variation; the same genetic basis was common to both.

Johannsen's work with pure lines showed that the variation due to the environment was not inherited. Only differences due to the genotype were passed on to the next generation. This was further strong evidence against the inheritance of acquired characters, but Johannsen thought that it also told against natural selection. When natural selection acts on continuous variation, it cannot shift the nature of the genotypes; it can only prefer one genotype to another within the population. Where there are no differences in the genotype, selection can have no hereditary influence.

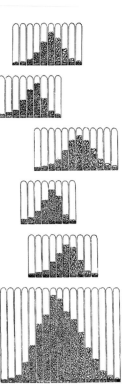

Johannsen's pure lines in the garden bean Phaseolus. A diagram showing five different pure lines and a population (BOTTOM) formed by mixing them together. The beans are arranged in equal classes of bean length, with each size class enclosed in a glass tube. Identical size classes are aligned vertically on the page. The separate lines differ in the mean value of bean length but, when mixed together, they produce a population with a smooth curve of variation.

Hence selection is dependent on mutation to make new genetic material available if it is to produce significant evolutionary change.

By mutation Johannsen meant the 'alteration, loss or gain of constituents of the genotype'. But to speak of mutations in this way, especially in the context of continuous variation, implied that they could be much smaller than the major discontinuities envisaged by de Vries. That this is indeed the case was demonstrated by Thomas Morgan, after he had begun to breed the fruit fly, *Drosophila*, in 1909. In one of his pure-bred cultures, a single white-eyed male appeared among the normal red-eyed flies. This white-eyed male was promptly bred with red-eyed females, and their offspring checked for eye colour. Although the F_1 generation were all red-eyed, white-eyed males reappeared in F_2, showing that the white eye colour was a recessive allele. Hence the 'white-eye' character must have originated by a sudden change in the gene for red eye.

Morgan deliberately used de Vries' term 'mutation' for this origin of a new allele by a sudden change in a single gene. So the word mutation came to be associated with changes in the genotype on a much smaller scale than the changes de Vries had observed in *Oenothera*. In Morgan's 'fly room' at Columbia University, the *Drosophila* flies were bred in their hundreds of thousands. By scrutinizing these flies carefully, he and his colleagues soon found a steady stream of new mutations that occurred naturally in the cultured flies. The discovery that small-scale mutations crop up

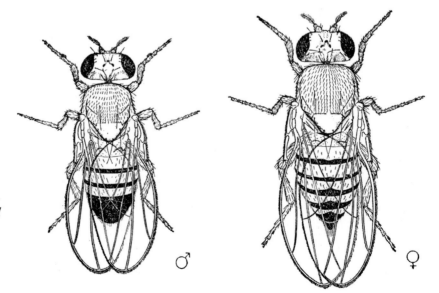

The fruit fly Drosophila. *Drawings of the male and female, from* The Physical Basis of Heredity *by Thomas Morgan.*

regularly under normal conditions was the key that enabled Morgan's team to unlock the mechanism of Mendelian heredity. It also paved the way for a fresh understanding of evolution in terms of natural selection.

With these new insights into the nature of continuous variation and mutation, the way became clear for a reconciliation between the views of the Darwinians and the Mendelians. Not that Bateson or de Vries accepted this; nor did Morgan himself, at first. Behind the scenes, however, there were others who had begun to see how the two views might be put together. A case in point was the British zoologist, Edwin Goodrich (1869–1946), based at Oxford. His own speciality was comparative anatomy, but he kept up with the new work in genetics and thought about its bearing on natural selection. As early as 1912 he published a popular little book, *The Evolution of Living Organisms*, which was enlarged and given a slightly different title in 1924.

There Goodrich explained that the role of variation in evolution could only be understood in the light of heredity. It was known, he wrote, that the hereditary substance is passed from one generation to the next in the nuclei of the germ cells, possibly in the chromosomes. The organism that developed from a fertilized egg could then vary due to two quite distinct causes: either through changes in the hereditary substance or through the influence of the environment during growth. Variations of the former kind were called 'mutations' and could be inherited by the next generation. The latter, which he preferred to call 'modifications' rather than 'acquired characters' or 'fluctuations', could not be inherited as far as one could tell.

The Mendelian 'laws of inheritance', worked out by crossing individuals that differ in some easily recognized characters, led to the conclusion that there was one hereditary unit for each character. Each unit or gene remained independent of the others, allowing different alleles to become segregated in the germ cells of a hybrid individual. All this work tended to confirm Weismann's view that bodily modifications are not inherited and so cannot contribute to the course of evolution. This left undirected mutations in the genes as the ultimate source of heritable variation. So far from weakening the case for natural selection, therefore, this new genetical research had 'definitely disposed of the only rival theory', that of the Lamarckians.

The new work also strengthened the case for natural selection, in Goodrich's view, by demonstrating that there is no hard and fast distinction between continuous and discontinuous variation. As Nilsson-Ehle's work showed, continuous variation comes about when a character is controlled by several genes, rather than just one. Natural selection could work effectively with the slight differences that are characteristic of continuous variation since they do have a genetic basis. And given a supply of mutations of the kind that Morgan was studying in *Drosophila*, natural selection could go on indefinitely accumulating changes in an adaptive direction.

In order to show that selection can indeed discriminate between small individual differences, Goodrich appealed to cases of field work. He cited the increase of a dark form of the peppered moth in the industrial north of England and the work of the zoologist Bumpus on the house sparrow in the Unites States. Bumpus had collected 136 sparrows that had been incapacitated by a snow storm; 72 of these revived but the rest did not. On measuring all the birds, he found that the survivors had, on average, a stockier build than those that died. Hence individuals that were better proportioned for withstanding the cold had survived, and the others had been eliminated. It was a clear example of natural selection acting on small individual differences.

This was about as far as one could take the discussion of natural selection and heredity at that time. Goodrich himself did not contribute to the further development of this discussion; he was preoccupied with his work in comparative anatomy and embryology. This line of work was at its peak about then, and it made a major contribution to the study of evolution, for it combined with the new fossil evidence to make evolution thoroughly convincing as an explanation for the history and diversity of life. Whatever doubts there were about the mechanism, there was no longer any doubt in the minds of biologists that evolution had occurred. This point was as firmly settled, Jordan and Kellogg said in *Evolution and Animal Life*, as the shape of the earth or the structure of the solar system: 'The earth is subspherical, the planets revolve about the sun, and species of organisms descend from other species.'

Those who studied vertebrate structure around the turn of the century could draw on a rich inheritance of careful studies going back to Cuvier and beyond. This enabled relationships and homologies to be worked out in considerable detail, not only for the skeleton but also for other parts of the body. Where some point of comparison was confusing in the adult forms, it often became clear when the embryonic forms were examined. This point-by-point comparison of structure yielded a consistent picture, supporting the view that all vertebrates are descended from a common ancestor. From this information, it was also possible to trace the probable course of vertebrate evolution.

In Haeckel's words, adults, embryos and fossils constituted 'three ancestral documents', from which the course of evolution could be inferred. While accepting this, Goodrich and his contemporaries were careful to distance themselves from Haeckel's views on recapitulation. It could no longer be supposed that embryonic development recapitulates the adult stages of ancestral species. The gill slits that occur in a mammalian embryo, for example, obviously correspond to the gill slits of embryonic fish and not the gills of adult fish. Many other structures characteristic of fish are not recapitulated at all. The truth of the matter seemed to be that the embryonic stages of some ancestral structures are retained simply as formative stimuli in development. This meant that embryos could provide valuable clues to evolutionary relationships, but they could not give definite information about the adult forms of ancestors.

It would be hard to find a more elegant reading of Haeckel's three ancestral documents than that concerned with the vertebrate head. The structure of the head in an adult vertebrate is complicated by the presence of the mouth and special sense organs (nose, eye and ear) as well as the protective skull. As a result of these features, the head is sharply differentiated from the rest of the body, especially in the more complex vertebrates. Nevertheless, if the theory of evolution is correct, this condition must have arisen from a simpler one where the head differed but little from the rest of the body. Once a distinct head had arisen, its features must have been progressively modified in the course of evolution from fish to reptile to mammal.

Toward the end of the nineteenth century, this problem was tackled with a segmental theory of the head. It was already known that the rest of the body in a vertebrate is built up from a row of repeated modules or segments. This segmentation largely disappears in the complex organization of the adult, although it is evident in the arrangement of the ribs, vertebrae and spinal nerves. But there is no trace of segmentation in the head of an adult vertebrate.

However, when studies were made of development in the early embryo of several species, it was found that the head is indeed composed of a number of segments. At this early stage, the developing head forms part of a complete row of segments that runs from the front to the hind end of the body. Each segment could be recognized primarily by a block of developing muscle, called a somite, to which a number was assigned beginning from the front.

Above each somite there could be found a segmental nerve, issuing from the developing brain and divided into distinct dorsal and ventral nerve roots. The ventral root innervates the somite, while the dorsal root supplies other structures within its segment. The gill slits develop below the level of the somites in the head region, with each slit separated from the next by a bar of tissue, the gill arch. A large branch of a dorsal nerve root was found to pass down each of these gill arches, showing that they correspond to the segmentation of the body. A supporting element of cartilage also develops within each gill arch.

All this went to show that the whole body of a vertebrate develops from a common segmental plan. Complex though it is, the head develops from segments like those in the rest of the body. The idea of vertebrates having a common ancestor that was a simple, uniformly segmented animal was thus upheld. A detailed paper on the development of head segments was

Goodrich's diagram of the head segments in a dogfish embryo. Each segment is distinguished by a muscle block, which is cross-hatched and given a number. The nerves are black, and skeletal elements of cartilage are shown by a dotted outline. The gill slits are identified by Roman numerals.

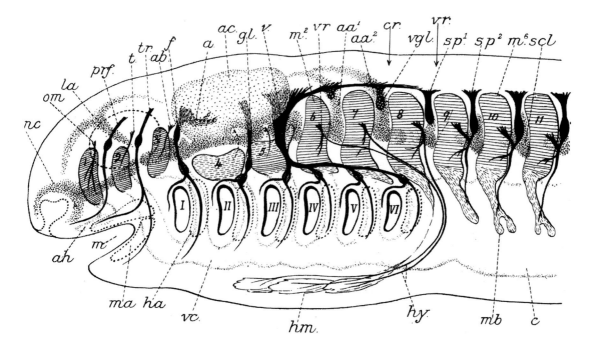

published by Goodrich in 1918, and he went on to publish a book on vertebrate structure and development in 1930. One topic that had a prominent place in this book was the evolution of the vertebarte head, as reflected in the bones of the skull.

Research on this topic was making swift progress at that time, mainly through examining the pattern of bones in fossil skulls. It was possible to make out the details because a join line or suture occurs where two bones meet in development, and this can still be seen in well-preserved fossils. As new and more complete fossil material was described, it became possible to trace the sequence of evolution from one class of vertebrates to another with increasing confidence.

What emerged was that a consistent arrangement of the outer bones was to be found in the skulls of early land-dwelling vertebrates. This bony pattern, which was similar to that in living reptiles and mammals, was also found in bony fish of an even earlier period. The main elements of this pattern included a ring of about five bones around each eye socket or orbit. Below this, there were a couple of prominent bones along the lower edge of the skull that bore the teeth. The roof of the skull was formed from some large, paired bones that met in the mid-line. And there were a variable number of 'temporal' bones on the side of the head behind the orbit.

The earliest fossils in which such a pattern of skull bones could be discerned were the bony fishes of the middle Devonian period. Even at this early stage of their history, the bony fishes were divided into two groups, the ray-finned fishes and the lobe-finned fishes. Whereas the former group had fins supported by parallel rays, the latter had fins supported by a central line of bones. The distinction between the two groups had been recognized since the middle of the nineteenth century when Agassiz had conducted his detailed study of fossil fish. Later zoologists saw that the arrangement of bones in the lobe-fins could have been the starting point for the evolution of the limb bones in land-dwelling vertebrates. Such a thought was made even more likely by the arrangement of bones in the skull.

The best known of the lobe-finned fishes was *Osteolepis*, several species of which had been described by Agassiz. When later workers came to study the skull of *Osteolepis*, they were able to pick out a number of bones that were homologous with those in the skulls of land-dwelling vertebrates. This was particularly true for the bones around the orbit and those further

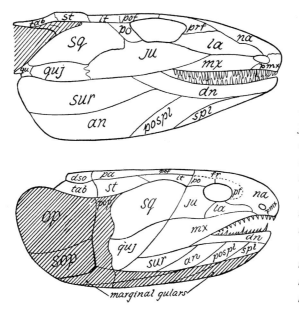

Skulls of a lobe-finned fish (BELOW) and an early amphibian (ABOVE). Published in 1929, these drawings were made to show the pattern of the skull bones. Corresponding bones in the two skulls are labelled with the same letters.

back on the skull. Right at the front of the skull there was a rather confusing arrangement of small bones, which were sometimes fused into a protective shield. In general, though, the homologies were clear.

The first vertebrates to leave the water were the early amphibians, which had been studied by Richard Owen. Over the years, the remains of a variety of these amphibians were dug from rocks around the world, mostly dating from the Carboniferous and Permian periods. The fossil finds were dominated by skulls, which evidently belonged to flat, heavily built heads somewhat like those of crocodiles. These skulls were found to be completely roofed with bone, with significant openings only for the two eyes and two nostrils.

They resembled the skulls of lobe-finned fish in this respect but differed in their proportions. In the fish, the part of the skull in front of the eyes was relatively short and the part behind the eyes was relatively long. These proportions were reversed in the early amphibians. The skull in front of the eyes was enlarged while that behind was shortened, and the relative sizes of the bones in the two regions were changed accordingly. Apart from the loss of some bones in the amphibians, the bony pattern was similar in the two groups of fossils. In view of this and the other similarities, it came to be generally accepted that the first amphibians had evolved from the lobe-finned fish.

Some remarkably complete and well-preserved fossils came to light as suitable deposits continued to be discovered and explored. Among these finds, the early land-dwelling vertebrates described and named by Cope were of particular interest. One of these, which he named *Eryops*, was an amphibian collected from rocks of the early Permian period in Texas. The whole skeleton was preserved and this enabled its anatomy to be examined and compared in minute detail. The mounted skeleton also made a striking museum exhibit and photographs of it began to appear in books early in the twentieth century.

Eryops was a strongly built animal with a massive skull and was about 1.5 m in length. The skull was broad and flat, with large openings for the eyes, as was usual among the early amphibians. The body behind the skull was supported by a remarkably strong vertebral column bearing a well-developed set of ribs. The skeleton also had two pairs of short and powerful limbs with broad feet, which extended out on either side of the body. This combination of features shows that *Eryops* was well adapted for life on land, although it probably moved with a rather lumbering gait.

Another detailed find from the Permian rocks of Texas was an early reptile, and this was named *Dimetrodon* by Cope. *Dimetrodon* and similar finds revealed what the early reptiles had looked like, before the arrival of the dinosaurs. From the structure of this animal's backbone and limbs,

Mounted skeleton of a fossil amphibian named Eryops *by Edward Cope.* Eryops *lived in North America during the Permian period and was fairly typical of the early amphibians.*

it was clearly more agile than the lumbering amphibians. It also had a deeper and narrower skull, which was an adaptation for accommodating long, strong jaw muscles that were capable of delivering a powerful bite. Along with this, there was an opening on the side of the skull behind the eye, which gave room for the bulging of the jaw muscles. The fact that there was only one opening showed that *Dimetrodon* was not related to the dinosaurs, which always had two such openings. Instead, it was related to another group that came to be known as the mammal-like reptiles.

The story of these animals began, once again, with Richard Owen. He recognized some mammal-like features in the teeth and jaws of fossil reptiles that had been shipped back to London from the site of British military road works in South Africa. These remarkable fossils, which were dug from the Karroo beds, dated mainly from the late Permian and the Triassic periods, later in time than *Dimetrodon*. From this rich source, dozens of genera and species were named by Owen and others. But the person who sorted out their place in evolution was Robert Broom (1866–1951), an energetic Scotsman who moved to South Africa in 1897. By about 1920, he had assembled a series of fossil forms that showed how this group of reptiles had gradually evolved into mammals.

Certain of these reptiles became more and more like mammals in several features of the skeleton that reflect an active, carnivorous way of life. The legs were generally tucked in under the body, with the elbows pointing backward and the knees forward. This is a pose that raises the body off the ground and so increases the efficiency of locomotion in four-footed animals. Differentiation of the teeth, which was first seen in *Dimetrodon*, progressed to the extent that distinct incisors, canines and molars made their appearance in the later forms. Such an array of teeth is adapted to cut up and crush food, which can then be digested rapidly, enabling a high level of activity to be sustained. The opening on the side of the skull also became enlarged, making room for even larger jaw muscles, and the dentary bone came to make up a progressively larger proportion of the lower jaw.

In most reptiles, the lower jaw consists of a number of separate bones in addition to the dentary, which carries the teeth. In the later mammal-like reptiles, the dentary formed almost the whole of the lower jaw and the other bones were small and crowded together. This made for a stronger jaw, and it resembled the mammalian condition in which the dentary is the only bone in the jaw. As in other reptiles, the jaw joint was between the articular bone of the lower jaw and the quadrate bone of the skull, but these bones were much reduced. Consequently, the dentary bone of the lower jaw and the squamosal bone of the skull, which form the jaw joint in mammals, were almost touching each other. Hence these reptiles were but a short step away from the full mammalian condition, as Broom clearly perceived.

This fossil evidence left one important point unresolved concerning the transition from reptile to mammal. Although mammals have fewer bones in the lower jaw than reptiles, they have two more in the middle ear, which is close to the jaw joint. In addition to the stapes, which is present in amphibians and reptiles, mammals have the incus and malleus making up a chain that links the auditory capsule to the ear drum. Could these

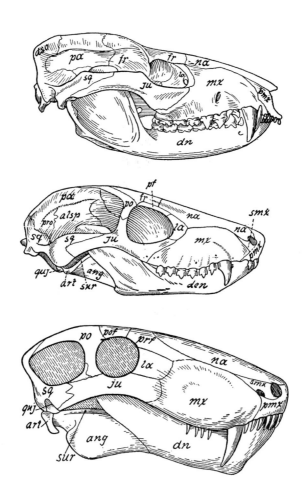

Skulls of an earlier mammal-like reptile (BELOW), a later mammal-like reptile (CENTRE) and a modern opossum (ABOVE). These drawings show the gradual shift from one form to the next in the pattern of bones in the skull and lower jaw. Corresponding bones in all three skulls are labelled with the same letters.

additional bones be derived from the now superfluous bones of the reptilian jaw?

Here the embryological work came into its own by providing strong evidence that this is indeed the case. It was shown that the incus and malleus develop from the same cartilaginous elements in the second or mandibular segment of the head as do the quadrate and articular bones in reptiles. These homologies, between the incus and quadrate and between the malleus and articular, were established through point-by-point comparisons of both adults and embryos.

Through the comparative study of living and fossil material, then, the broad outline of vertebrate evolution had been made clear by the 1920s. Good fossil evidence had been unearthed for the transitions from fish to amphibians to reptiles to mammals. This new evidence was often featured prominently in books on evolution or vertebrate zoology that appeared in the early twentieth century. However, it would be a mistake to think that evolution was seen as a single unfolding sequence leading to humans.

On the contrary, the diversity of evolution at all levels was clearly appreciated, and had been emphasized since the turn of the century by Henry Osborn in particular. Osborn (1857–1935), who had trained under Cope, was keenly interested in the pattern of vertebrate evolution that could be read from the fossil record. He noticed the tendency for a new group of animals to diversify rapidly after its first appearance in some generalized form. The new group soon evolved into a variety of specialized forms that represented adaptations to the major world environments and to different ways of obtaining food. For this pattern of evolution he coined the term 'adaptive radiation', and it caught on at once.

Such a pattern was illustrated not only by the mammals, which Osborn had discussed, but also by the reptiles and by major groups of invertebrates. Adaptive radiation was also evident at lower levels of diversity. For instance, the hoofed mammals (ungulates) first appeared as a small group of unspecialized forms, the condylarths, to which Cope's genus *Phenacodus* belonged (see illustration on page 141). Subsequent evolution gave rise to a whole range of hoofed mammals, not all of which have survived to the present.

Darwin would have been delighted with this new understanding of vertebrate history, for he had despaired of the fossil record. In the *Origin*, he described it as 'a history of the world imperfectly kept', of which only a few scattered and incomplete pages have come down

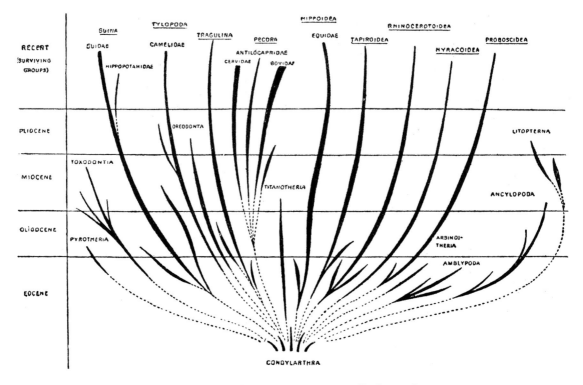

to us. Gratifying though the new knowledge was, it was really limited to evolution on a broad scale. The finer scale of events, involving the transformation of one species into another, was still pretty much a mystery. However, this was about to change. Over the next two or three decades, new work in genetics provided an interpretation of natural selection that could be reconciled with the work of field naturalists on species. The result was a new synthesis of evolutionary theory.

Evolution of the hoofed mammals. A diagram drawn to emphasize their adaptive radiation from an original unspecialized group, the condylarths, in the Eocene. From Goodrich's Evolution of Living Organisms.

CHAPTER EIGHT

SYNTHESIS AND SPECIES

Information published in a book does not always have the effect that the author intended. Occasionally, a book may even carry information that proves to be the seed of its own destruction. This is what happened with a book on *Mimicry in Butterflies*, published in 1915 by R. C. Punnett, who was with Bateson at Cambridge. In describing the many cases of mimicry that were known by then, Punnett used them to support the views of Bateson and de Vries on evolution. He argued that mimetic resemblances could not have been forged gradually through natural selection but must have arisen at a single step through a large mutation. But one item in his book acted as a trigger for new research that would eventually sweep away this whole point of view.

The item in question was an appendix, which contained a table of figures prepared by the Cambridge mathematician H. T. J. Norton. This table provided an estimate of the number of generations it would take for a new mutation to spread through a population under the influence of selection. It was assumed that the character controlled by this gene showed complete dominance, so that the heterozygote could not be distinguished from the dominant homozygote. Separate figures were then given for where the new mutation was dominant and for where it was recessive. In each case, the result was calculated for four different intensities of selection, ranging from a 50% down to a 1% selection intensity.

These estimates drew on calculations made some years previously by G. H. Hardy in the United States. He had worked out how a pair of alleles, *A* and *a*, would be distributed in a large population, in which there was no selection operating. All the members of the population must belong to one of three genotypes, *AA*, *Aa* and *aa*, and through Mendelian segregation these will yield offspring in constant proportions. Then, if the alleles *A* and *a* are initially present in the ratio $p:q$, where $p+q=1$, it could

Percentage of total population formed by old variety	Percentage of total population formed by the hybrids	Percentage of total population formed by the new variety	Number of generations taken to pass from one position to another as indicated in the percentages of different individuals in left-hand column							
			A. Where the new variety is dominant				B. Where the new variety is recessive			
			$\frac{100}{50}$	$\frac{100}{75}$	$\frac{100}{90}$	$\frac{100}{99}$	$\frac{100}{50}$	$\frac{100}{75}$	$\frac{100}{90}$	$\frac{100}{99}$
99·9	·09	·000								
98·0	1·96	·008	4	10	28	300	1920	5740	17,200	189,092
90·7	9·0	·03	2	5	15	165	85	250	744	8,160
69·0	27·7	2·8	2	4	14	153	18	51	149	1,615
44·4	44·4	11·1	2	4	12	121	5	13	36	389
25·	50·	25·	2	4	12	119	2	6	16	169
11·1	44·4	44·4	4	8	18	171	2	4	11	118
2·8	27·7	69·0	10	17	40	393	2	4	11	120
·03	9·0	90·7	36	68	166	1,632	2	6	14	152
·008	1·96	98·0	170	333	827	8,243	2	6	16	165
·000	·09	99·9	3840	7653	19,111	191,002	4	10	28	299

be shown that the frequency of *AA*, *Aa* and *aa* in the next generation will be $p^2 : 2pq : q^2$. Hardy went on to show that this ratio will not change in subsequent generations. In other words, the two alleles, *A* and *a*, will remain in the same proportions in a population from generation to generation unless some disturbance such as natural selection upsets the equilibrium.

When Hardy's paper was published in 1908, it laid the foundation for the study of genes in populations, and Norton was one of the first to take advantage of this. By taking a Mendelian pattern of inheritance for granted, mathematics could be used to estimate what would happen to a given allele in a population through successive generations. Norton's table demonstrated that natural selection had the power to change the frequency of a gene within a population. Even a slight advantage, of 10% or less, would lead to a substantial increase in the frequency of a favourable gene over relatively few generations. This finding came as a surprise to most people, and it stimulated a flow of new research.

In Cambridge, the mathematical approach to selection theory was taken up by Ronald Fisher (1890–1962), who had a keen interest in evolution. Beginning in 1918, he published a series of papers on the distribution of genes in populations, using statistical techniques to study the effects of selection, mutation and genetic variation. He was able to show that a Mendelian pattern of inheritance is entirely consistent with a theory of gradual evolution based on the natural selection of small genetic differences. By 1927, when he published a direct reply to Punnett's book on mimicry, he felt confident that the day of large mutations in evolution

Norton's table giving estimates of the rate at which a gene could increase in a population through selection, from Punnet's Mimicry in Butterflies. In modern terms, 'variety' means allele or gene and 'hybrid' means heterozygote.

was over. 'It is now becoming increasingly widely understood,' he wrote, 'that the bearing of genetical discoveries . . . upon evolutionary theory is quite other than that which the pioneers of Mendelism originally took it to be.'

The new outlook was made generally available with the appearance of Fisher's *Genetical Theory of Natural Selection* in 1930. There he pointed out that with Mendelian inheritance there is no inherent tendency for variability to diminish with time. As Hardy had shown, the proportions in which a pair of alleles are found does not change in the course of normal reproduction. Alternative genes are therefore conserved in a population unless their proportions are changed by chance, by mutation or by selection. Fisher's calculations showed that selection is by far the most effective of these factors that can change gene frequency. Chance survival or loss of genes could have a significant effect only in small, isolated populations, and mutation was too rare to have much effect on gene frequency. It was known from *Drosophila* that a given gene seldom mutates in more than one individual in 100,000.

This being so, all those theories that assumed mutations were the driving force of evolution must be set aside. Any hypothetical agency that could control the direction of mutations would be quite ineffective in controlling the direction of evolutionary change. Even a minute selective disadvantage would be able to overpower it. Nor could this conclusion be avoided by supposing that Mendel's laws did not apply to some cases of inheritance, for such doubtful cases were now yielding to a Mendelian interpretation. The continuous variation in human height, for example, could be accounted for by the action of several genes that are additive in their effects. The degree of correlation found in the height of relatives was exactly what one would expect from a series of genes showing normal Mendelian dominance.

Hence natural selection was the only theory of evolutionary change that was compatible with the new discoveries in genetics, contrary to what had been thought earlier. Much the same conclusion was reached independently by another British biologist, J. B. S. Haldane, who had been moved to do similar calculations after seeing Norton's figures. From 1924 onwards, he produced a series of papers on 'A mathematical theory of natural and artificial selection'. In the first of these, he applied his computations to an actual case, the spread of the black form of the peppered moth, *Biston betularia*, in industrial areas. A black form of this moth had first been recorded near Manchester in 1848, and by 1895 it had almost completely replaced the normal peppered grey form in that area.

Breeding experiments had shown that the black form of the peppered moth was due to a single dominant gene. Haldane worked out that, for this gene to spread through the population in so short a time, the black form must have had a 30% selective advantage over the grey form. This was an early indication that selection intensities in nature might be much more than the figure of around 1% envisaged by Fisher. Even this lower figure was quite sufficient to ensure that the selected form would eventually spread through the population, as Fisher showed.

In following through the mathematical approach to the spread of genes in populations, one of Fisher's most original contributions was his treatment of 'polymorphism'. This is the name given to a situation where two or more discontinuous forms of a species occur together in the same region. Fisher demonstrated that selection might act to maintain a balance between two alternative alleles in a population under certain conditions. For instance, one of the alleles might have a selective advantage only up to a certain frequency, above which it became disadvantageous. Again, if the heterozygote should happen to have a selective advantage over both homozygotes, a stable equilibrium between the two alleles would result. Interactions such as these would bring about a balanced polymorphism, in which alternative forms were maintained in the species for long periods. As a result, natural populations might be expected to contain a wide range of genetic variation.

Another piece of evidence that pointed in the same direction was the effect of interaction between separate alleles. Two genes, represented by A, a and B, b, could maintain each other in equilibrium if A was advantageous in the presence of B but disadvantageous in the presence of b, and vice versa for B. That genes might interact in this sort of way was clear from work done with real animals and plants. For while Fisher and Haldane were busy with paper and pencil, other workers were conducting breeding experiments that threw light on the theory of natural selection. The most rapid progress in genetics was being made by Morgan and his colleagues on *Drosophila*, but significant work was also being done on vertebrates.

One person who carried out an influential series of breeding experiments was William Castle (1867—1962) in the United States. One of his earlier experiments involved a direct test of whether the phenotypic characters of vertebrates can influence the genes that are passed on to the next generation. He transplanted the ovaries of a black guinea pig into an albino female whose own ovaries had been removed. This albino female was then mated to an albino male, but in three successive litters she bore only black offspring. When an account of this work was published in 1911, it was another blow to the idea of the inheritance of acquired characters. Subsequently, Castle provided strong evidence for the efficacy of selection and for gene interactions through breeding experiments with rats of different coat colours.

He was able to show that the 'hooded', black and white pattern behaves as a simple Mendelian recessive to the normal grey of wild rats. The hooded pattern was quite variable, and through continued selection it was possible to obtain strains of hooded rats that were almost entirely black or almost entirely white. These two strains were quite stable, for return selection to the typical pattern was no easier than the original selection. When the two selected strains were each back-crossed with homozygous grey rats, the hooded patterns still behaved as simple recessives. The F_1 generation were grey heterozygotes, and by inbreeding these the hooded pattern could be recovered in F_2.

However, the hooded rats in F_2 had less-developed patterns than their hooded grandparents: the almost-black rats had less black and the almost-white ones had less white. Hence the departures from the typical hooded pattern must have been achieved by selecting a combination of other genes that modified the effect of the hooded gene. If the modification had taken place in the hooded gene itself, the hooded rats in F_2 should have received fully modified hooded genes and so been the same as their grandparents. As it was, the appearance of the F_2 hooded rats was evidently due to their

The range of coat patterns obtained in 'hooded' rats by Castle and Phillips. The normal piebald pattern is indicated by the value 0 on the scale above, and positive and negative values indicate patterns resulting from selection for greater or lesser areas of black, respectively.

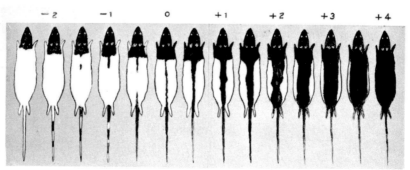

having inherited selected and unselected modifiers equally from the two sets of grandparents.

This result was one of the first to show that interactions between genes are important in contributing to the final appearance of the phenotype. A single gene does not control a unit character entirely on its own, but its effects can be modified by the action of other genes. Looked at another way, this means that individual genes are likely to influence more than one aspect of the phenotype. This effect was demonstrated particularly well in the *Drosophila* work, and was called pleiotropy. A consequence of pleiotropy is that each organism tends to be a little different from the next, according to the particular combination of genes that are carried. This can provide a large store of variation for natural selection to work upon, as Castle realized in his 1914 report on the hooded rats. If these gene interactions really occur, he wrote, then 'selection is an agency of real creative power, able to modify unit characters indefinitely so long as physiological limitations are not reached'.

By the time this series of back-crosses was completed in 1919, Castle was sure that genes really can modify the effects of other genes. The idea of making these crucial back-crosses had come from a student of his by the name of Sewall Wright. As well as participating in the experiments with hooded rats, Wright (1889−1988) made a detailed study of the inheritance of coat colour in guinea pigs during his time in Castle's laboratory. On moving to the US Department of Agriculture, he took charge of a long-running experiment on inbreeding in guinea pigs and developed new methods to analyse the accumulated results. All this experimental work persuaded Wright not only that selection was effective but also that it was most effective with groups of interacting genes. From his experience, he was convinced that interactions between genes are an important part of the hereditary make-up of living organisms.

With the inbreeding experiment, he found that the inbred families became different from each other as the generations passed and they became more and more homozygous. Evidently, genes from the original heterozygous stock were lost through inbreeding, and it was a matter of chance as to which combinations of genes became fixed in different families. It occurred to him that inbreeding could be used with livestock to generate lines in which different combinations of characters became fixed by chance. Then those lines with desirable combinations could be crossed to restore the vigour that was usually lost during inbreeding.

The resulting crossbred stock would contain a combination of desired characters that might never have been obtained by selecting single characters alone.

A similar set of circumstances, Wright believed, would lead to the appearance of new genetic combinations in evolution. Natural selection should work most effectively in smaller populations, where significant errors of sampling would occur. Just as a small sample may give us an erroneous idea of product quality or popular opinion, so a small population may contain a selection of genes that is unrepresentative of the species as a whole. Wright considered that the flow of genes from adjacent populations would usually prevent genes being fixed or lost by chance. Nevertheless, there would be sufficient sampling error or 'genetic drift' to permit gene combinations that would be unlikely to occur in larger populations. Natural selection would then act upon these new combinations, picking out those with the best gene interactions, and the species would evolve more rapidly than by the selection of single genes.

In adopting this view, Wright was offering a somewhat different interpretation of the evolutionary process from that of Fisher and Haldane. The latter both believed that natural selection would act most effectively on single genes in large populations, which have a greater store of genetic variation. When his first major paper on evolution was published in 1931, Wright's idea of genetic drift caused some dispute, along with other points of disagreement.

In spite of these differences, and their differing mathematical methods, Fisher, Haldane and Wright were agreed on a number of basic conclusions. All three recognized the power of natural selection to change gene frequencies in populations and the inability of mutations to produce such changes on their own. A consequence of this, they all agreed, was that evolution must take place gradually, through the selective accumulation of small genetic differences, contrary to the views of Bateson and de Vries. And they all confirmed that the continuous variation shown by many individual differences is not in conflict with the underlying discontinuity of genes.

These conclusions did much to bridge the gap that existed between the geneticists in their laboratories and the naturalists who worked in the field. For a time, these two groups had been so different in their approaches to evolution that it seemed as if they could never come to a common mind.

It will be recalled that this situation had come about chiefly through the claim of the early Mendelians that genetic variation is discontinuous. They thought that a new species would arise through a single drastic mutation, whereas naturalists were convinced that most variation is continuous and that the formation of a new species is a gradual process.

Naturalists had developed these views by studying the way that a species varies geographically when it is distributed over a large area. This study was not so much concerned with distribution on a global scale, but with geographical variation within one species or a local group of species. Back in the eighteenth century, both Linnaeus and Buffon had been aware that populations in different parts of a species' range may be slightly different from each other. The most obvious explanation for this seemed to be that the differing populations were exposed to somewhat different climates in their respective localities.

Naturalists began to give this topic more attention, and as early as 1833 a whole book was devoted to *The Variation of Birds Under the Influence of Climate* by C. L. Gloger. He showed that the members of a species living in a more humid climate tend to have darker plumage than those in a drier climate, and this became known as 'Gloger's rule'. Further correlations with climate were found, especially in mammals, and were widely assumed to be adaptations to local conditions.

As naturalists travelled the world, it was noticed that differences within a species were accentuated by partial barriers to free migration. The result was that a species became divided into a number of 'geographical races' or 'subspecies', which lived in different localities and were distinguished by slightly different anatomical features. Often it was difficult to tell whether these races should be ranked as separate species or only as local varieties within a species. This kind of situation was described by Wallace in butterflies from the islands of South-East Asia and by Bates in those he found along the tributaries of the Amazon River. 'We seem to obtain here a glimpse of the manufacture of new species in nature', Bates remarked in the book of his travels. Although he and Wallace saw this kind of geographical variation as evidence for evolution, they did not think out in detail how a new species might split off from an old one.

Darwin gave some attention to this point in his big book on *Natural Selection*, and there he distinguished clearly between two different classes of evolutionary change. One is where a species is slowly modified as it becomes adapted to changing conditions in its locality, and the other is where a species splits into two or more new species. The first case is a matter of progressive adaptation, while the second is a matter of increasing diversity. In the latter case, Darwin thought that 'some degree of isolation would generally be almost indispensable' for the formation of new species. For without isolation, the free interbreeding of populations that varied in different ways would tend to restore uniformity and so prevent selection from adapting them in divergent directions.

The main way in which populations could become isolated was by geographical separation, through some barrier to migration. Darwin decided that the ideal situation for producing new species would be a large tract of land temporarily converted into a group of islands through changes in sea level. However, when he came to condense this argument for the *Origin of Species*, Darwin did not stress the role of geographical isolation in species formation. In fact, it is not at all clear in the *Origin* that he attached any importance to it. But he did make it clear that local varieties or subspecies had the potential to become new species; he regarded them as 'incipient species'.

After the *Origin* was published, the question of isolation was taken up enthusiastically by the German naturalist and explorer Moritz Wagner.

A diagram of species formation from Darwin's Natural Selection. *A to M represent species in a genus of plants becoming adapted to wet terrain (to the left) or dry (to the right). As it evolves, species A splits into three new species, while species M is progressively modified without splitting.*

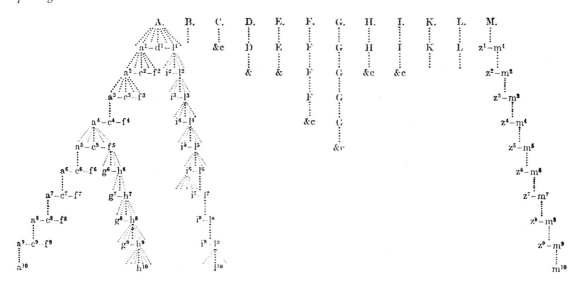

On his travels, Wagner (1813–87) had seen for himself how distinct but closely related species are often separated by barriers such as rivers or mountains. What this pattern implied, he felt sure, was that geographical isolation is an essential part of the process of species formation. A new species could be formed only when some individuals crossed the previous limits of their range and become separated for a long time from the other members of their species. Only long continued separation from other members of the species could provide the right condition for natural selection to turn a population into a new and separate species. Without this isolation, the work of natural selection would always be undone by free interbreeding within the species.

Not content with having drawn attention to an important point, Wagner went on to promote migration and isolation as a complete alternative to Darwin's theory. He even tried to explain adaptation through migration, supposing for example that polar bears were descended from albino individuals that had migrated to the snow to boost their chances of survival. Such views moved Darwin to write the words 'Most Wretched Rubbish' across the front of his copy of a paper by Wagner in 1875. Despite his excesses, Wagner's views did help to bring out the importance of isolation in the formation of new species. By the end of the nineteenth century, it was widely agreed that a new species cannot be produced without the isolation of a population. Not all naturalists agreed that spatial separation was essential; some thought that an incipient species could be isolated just as well by new habits or ecological preferences.

One person who took the latter view was John Gulick (1832–1923), who discovered what was to become a classic piece of evidence for the importance of geographical isolation. He had made a special study of the land snails of the Hawaiian island of Oahu, where he found that the species of snails in the genus *Achatinella* are extraordinarily localized. In most cases, each species is confined to a single valley no more than three or four miles in extent. So a genus of snails is represented in successive valleys, which are separated by narrow mountain spurs, by a succession of separate species. He found a strong correlation between the extent to which species resemble each other and how far apart they are, with the more dissimilar ones being found in more widely separated localities (see colour plate 13).

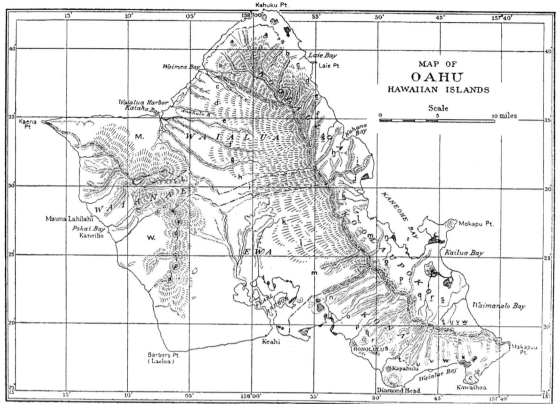

Gulick's map of the island of Oahu, on which the locations of snail species in the genus Achatinella *are indicated by small letters. For instance,* q, r *and* s *are small valleys near Honolulu, and the species of snail found in them are indicated by the corresponding letters in colour plate 13.*

In fact, each species was found to have even more localized races that not only grade into each other but also link one species to the next. The most divergent species on the island are therefore connected by a continuous series of intermediate forms. This meant that the dividing line between neighbouring species was rather doubtful. Indeed, it was difficult to tell just how many species there really were, and workers who came after Gulick settled for a much smaller number of named species.

Gulick naturally came to the conclusion that it was the isolating effect of the mountain spurs that had permitted a closely related group of snails to become so diversified in so small an area. The graded series of forms along the valleys strongly suggested that all had arisen from one original species through a consecutive series of isolations. In turn, this implied that isolation was an essential factor in the evolution of divergent forms, through races to species. Natural selection could not turn one species into two or more unless free interbreeding between the evolving forms was prevented by their isolation. By the time Gulick came to summarize his life's work in 1905, other naturalists were coming to much the same conclusion from the study of birds and mammals.

The species of these larger and more mobile animals exhibited a similar pattern of distribution on a larger scale. In these animals also, the species most alike in structure were not found in the same locality, nor far apart, but in neighbouring localities. Where the two neighbouring forms were found not to intergrade, they were classified as distinct species. If there was intergradation, with the two forms being linked by individuals of intermediate structure, each of the connected forms was called a subspecies. It became customary among those who classified birds and mammals to recognize a subspecies formally by adding a third name after the two Latin names in the Linnaean nomenclature.

In 1909, for example, the US Department of Agriculture published a classification of native mice belonging to the genus *Peromyscus* by W. H. Osgood. One of the species that he revised, *Peromyscus maniculatus*, is found all over North America and has many geographical races, which grade into each other (see colour plate 14). Each of these he classified as a subspecies, using a third name. So the race found in southern California was named *Peromyscus maniculatus gambeli*, that found in the Mojave Desert was named *Peromyscus maniculatus sonoriensis*, and so on (for convenience, such names are usually shortened to *P. m. gambeli* and *P. m. sonoriensis*, respectively, after first being given in full). Where there was no sign of intergradation, the local form was named as a distinct species, as in the case of *Peromyscus polionotus*, which is found in Alabama and Florida.

Field workers, who were familiar with this sort of situation, had little time for the notions of de Vries and Bateson that new species might originate through single mutations. But the gap between these contrasting points of view was certainly reduced by the changing perspective of the geneticists. The gap was also bridged from the other side by those who decided to apply genetical analysis to the populations they studied in the field. There is no better example of this approach than Francis Sumner in the United States. Sumner (1874–1945) had developed a strong interest in evolution as an undergraduate and longed to tackle some of the major problems with the aid of hard evidence. His opportunity came in 1913, when he began a study of the geographic races of *Peromyscus* at the Scripps Institution in California.

Sumner trapped large numbers of wild mice from the Californian races of *P. maniculatus* and began to measure them and to breed from them. He confirmed that the races differed, on average, in such measurable characters as total size, bodily proportions and coat colour. The differences between the races in the mean values of these characters were maintained through successive generations even when the mice were reared in a common environment. This showed that the distinctive features of the geographical races were genetically determined and were not just phenotypic differences due to the influence of their local environment. The individual differences within a race proved to be partly genetic and partly phenotypic, as one would expect, and were of the same kind as the racial differences. So there were darker and paler races, with darker and paler individuals within each race.

Sumner went on to cross different races of *Peromyscus maniculatus*, beginning with *P. m. gambeli* and *P. m. sonoriensis*, and paid special attention to coat colour. The F_1 generation was intermediate in appearance and seemed to be a blend of the two parental races. When these were bred among themselves, the F_2 generation had the same general appearance as F_1 but showed a somewhat greater range of variability. There was no sign of Mendelian ratios or dominance here, and the features of the two races appeared to be permanently blended. By contrast, Sumner found a number of distinctive colour 'mutants', such as albinism and yellow coat, that behaved as simple Mendelian recessives. He therefore fell in with the view that there are two quite different kinds of heredity: a Mendelian kind for sports and mutations, and a blended kind for continuous variation.

Although he knew of the multiple gene theory for explaining continuous variation in Mendelian terms, Sumner was sceptical about it at first. He was inclined to stick to the Lamarckian view that the graded differences between races were produced by the direct action of the environment. Coat colour, for example, could be determined by climate since mice with darker colours came from the more humid districts, in keeping with Gloger's rule. However, his views received a 'rude shock' when an extremely pale race, *Peromyscus polionotus leucocephalus*, was discovered in 1920 on an island close to the Florida coast, a region of high rainfall and humidity. This island race differed strikingly from the dark mainland race, *Peromyscus polionotus polionotus*, which occupied a large area of northern Florida and Alabama.

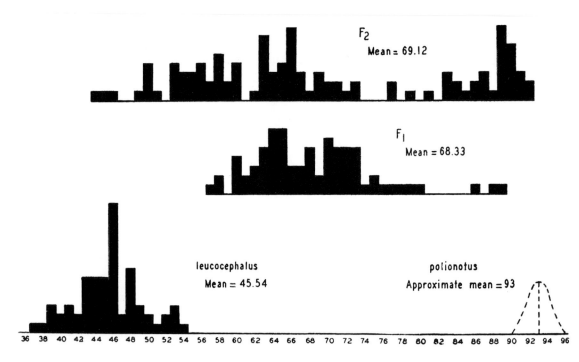

F₂ Mean = 69.12

F₁ Mean = 68.33

leucocephalus Mean = 45.54

polionotus Approximate mean = 93

36 38 40 42 44 46 48 50 52 54 56 58 60 62 64 66 68 70 72 74 76 78 80 82 84 86 88 90 92 94 96

This situation was an ideal one for analysis with the techniques that Sumner had perfected through years of work on *Peromyscus*. When he cross-bred mice of the *leucocephalus* and *polionotus* races, the coat colour of the F₁ generation was an intermediate blend between the two races. But then the F₂ generation showed a much wider range of variation, with the extreme cases having coat colours similar to those of their grand-parents (see colour plate 15). This was clear evidence of Mendelian segre-gation, and the results fitted the predictions of the multiple gene theory so exactly that Sumner now accepted it. He estimated that a minimum of five or six independent genes were involved in the differences in coat colour between the races.

There was a third race, *P. p. albifrons*, which was lighter than *polionotus* but darker than *leucocephalus* and was confined to the coast of southern Alabama and Florida. As well as crossing this race with the other two, Sumner also took a range of samples across the border between *albifrons* and *polionotus* to its north. He found that the one race graded smoothly into the other but that the region of intergradation was quite narrow and abrupt. The sampled populations changed from one set of racial char-acters to the other over a distance of about 10 miles. There was no such

The inheritance of coat colour in the mouse Peromyscus polionotus, *as analysed by Sumner. The histograms show the distribution of coat colour in the two races* leucocephalus *and* polionotus *(BELOW), in the F₁ generation when the two races are crossed (CENTRE) and in the F₂ generation (ABOVE). Coat colour is expressed in a graduated scale at the bottom, with paler to the left and darker to the right.*

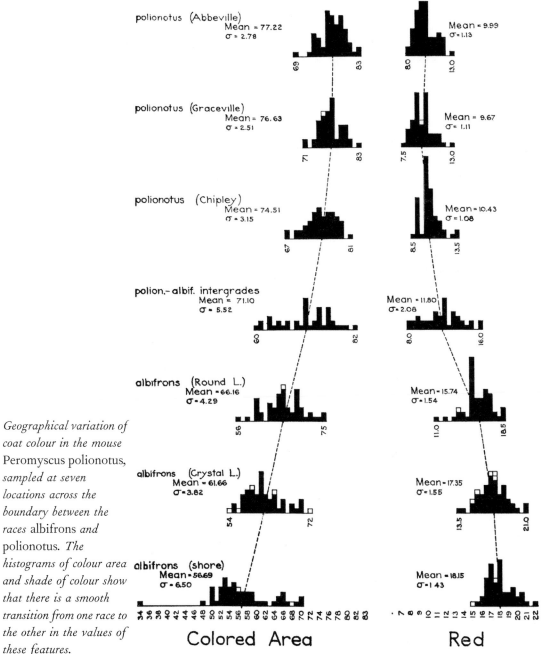

Geographical variation of coat colour in the mouse Peromyscus polionotus, *sampled at seven locations across the boundary between the races* albifrons *and* polionotus. *The histograms of colour area and shade of colour show that there is a smooth transition from one race to the other in the values of these features.*

Colored Area

Red

abrupt change in the climate, and Sumner conceded that the origin of the paler races, *albifrons* and *leucocephalus*, could be better explained in terms of protective coloration. The two paler races were well matched to the white sands on which they lived, and this could have been brought about by natural selection.

By the time he finished the work on *Peromyscus* in 1930, Sumner had reached some important conclusions on species and evolution. It was clear to him that the differences between subspecies are strongly inherited and depend on the action of many independent genes. From this result, he concluded, 'it seems certain that even the geographic subdivisions of a species do not arise as single mutational steps'. Indeed, there is so much genetic diversity in natural populations, due to mixing of the many genes in sexual reproduction, that a significant change could be brought about by selection in the entire absence of new mutations. Natural selection, in conjunction with some degree of geographical isolation, is therefore the main force causing the divergence of races, which have the potential to become new species through continued gradual evolution.

Such conclusions dovetailed nicely with those being drawn by the mathematical geneticists at about the same time. The importance of Sumner's work lay in the fact that he brought genetics and the geographical study of natural populations together. No one had done this before, and so his was the first detailed demonstration that the Mendelian principles worked out in the laboratory were applicable to populations of animals in the field. The next step in forging a synthesis of ideas owed much to an entirely separate group of researchers, but one in complete sympathy with Sumner's approach. This was the Russian school of evolutionary genetics, represented at large in the person of Theodosius Dobzhansky (1900–75).

In Russia, genetics had been closely linked to the work of naturalists from the first, and the importance of natural selection was taken for granted. During the 1920s, a group of geneticists with a keen interest in evolution had flourished under the leadership of Sergei Chetverikov at Moscow. Having started out as a naturalist, Chetverikov then took up genetics and made a special study of variability in wild populations of *Drosophila* around Moscow. He published a notable paper on evolution in the light of genetics in 1926, in which he came to conclusions remarkably similar to those of Fisher, Haldane and

Wright. This work was cut off by political events in 1929, but by then it had exerted a strong influence on Dobzhansky, who had been in Leningrad.

So when Dobzhansky joined Morgan's *Drosophila* team in the United States, he brought with him the evolutionary perspective of the Russian workers. With this background, he was in a good position to clear up the misunderstandings that had arisen between the field naturalists and the laboratory geneticists. This he did in a carefully thought-out book on *Genetics and the Origin of Species*, published in 1937, ten years after his arrival in the United States. In this book, Dobzhansky was able to bring together the results of field work, and of experimental and mathematical genetics, to form a single connected argument on the process of evolution. This was the first time that any such attempt had been made, and it was particularly well done. Consequently his book played a key role in influencing a wide range of people who had hitherto approached evolution from entirely separate points of view.

Dobzhansky began with what was known of mutation as the source of genetic variation, and drew on the wealth of information available from *Drosophila*. It was clear from this that mutations can alter a great variety of bodily characters, and that their effects range all the way from drastic changes to minute changes that are barely detectable. These mutations are evidently the source of genetic variation, which is stored in natural populations and forms the raw material for natural selection to act upon. This is true even when the characters in question are continuously variable, for these cases are due to the interaction of multiple genes, as Dobzhansky carefully explained. He pointed out that the graduated differences between geographical races could therefore have originated as numerous small mutations, citing cases such as Sumner's work.

In turning to the role of natural selection, Dobzhansky treated it not just as a mathematical theory but as a process that can be studied in natural populations. Among other things, he drew attention to changes in the composition of populations that had taken place in historical times, such as the spread of black forms of several species of moth. Changes of this kind could reasonably be attributed to the action of natural selection. An example that had recently come to light was

The geographical distribution of insecticide-resistant (black) and nonresistant (stippled) strains of the scale insect Aonidiella aurantii *in southern California. The map shows the spread of resistance through neighbouring districts, from Dobzhansky's* Genetics and the Origin of Species.

the spread of strains of scale insects that were resistant to the cyanide gas used to control them in the citrus groves of California. In some species of scale, resistant insects had first appeared in one locality, and resistance had then spread through neighbouring populations in the following years. Dobzhansky considered that this case 'constitutes probably the best proof of the effectiveness of natural selection yet obtained'.

Furthermore, the correlations that had been found between geographical variation and the local environment could be accounted for in terms of natural selection. Correlations such as Gloger's rule had often been interpreted in Lamarckian terms by naturalists, who had assumed they came about through the influence of the environment on the phenotype. But to suppose that geographical races had arisen through the inheritance of acquired characters was 'contrary to the whole sum of our knowledge' on genetics. Now that the genetic basis of continuous variation was known, it made much more sense to view these regularities in geographical variation as adaptations brought about through mutation and selection.

From here, Dobzhansky moved on to consider how new species are formed. He pointed out that races and species were known to differ from each other in many genes, which made it virtually impossible for a new species to arise through a single mutational step. The process of species formation must be a gradual one, in which distinct combinations of genes are built up over time. These gene combinations break down when races are interbred, and hence populations must be kept isolated for distinct gene combinations to form and be maintained. 'Species formation without isolation is impossible' was Dobzhansky's view.

In the early stages of species formation, geographical isolation could prevent two populations from interbreeding, but as a rule this would be only a temporary state of affairs. The separated populations might come together once more and would then interbreed unless physiological mechanisms that reduce or prevent interbreeding had evolved in the meantime. Hence a species is that stage in the evolutionary process 'at which the once actually or potentially interbreeding array of forms becomes segregated into two or more separate arrays which are physiologically incapable of interbreeding'. A species is therefore not an arbitrary level of classification but a natural unit, the 'biological species'.

In this unit, organisms are adapted to a particular way of life by possessing a particular combination of genes, which is protected from disruption by 'isolating mechanisms' that keep species apart. Dobzhansky gave a list of the sort of things that could act as isolating mechanisms. For instance, members of two species in the same area might be kept apart by being confined to different habitats or by breeding at different seasons. Alternatively, the two species might occur together but hybridization would be prevented by differences in breeding behaviour or by a simple inability to copulate. If differences like these evolved during the period of geographical isolation, they would keep two populations isolated even after the geographical barrier had broken down.

In developing this picture of species formation, Dobzhansky had taken into account the work of field naturalists. So it is not surprising that his views soon received support from naturalists, and none more so than Ernst Mayr, who had made a special study of geographic variation in birds. After several years of field work on the birds of New Guinea and the Solomon Islands, Mayr (1904–2005) moved from Germany to the United States in 1930.

At first, Mayr had been one of those naturalists who interpreted the facts of geographical variation in Lamarckian terms. But as he became acquainted with the new genetical work in the 1930s, he changed his mind and adopted the Darwinian explanation put forward by Dobzhansky. In 1942, he published *Systematics and the Origin of Species*, which summarized the field evidence on species formation and sought to link it with genetics.

The daily work of the systematist, Mayr explained, is to classify living things into species and higher categories. And in trying to place the organisms collected in the wild into their correct species, '90 percent of his work consists in the study of variation'. It had now been established not only that individuals of a species vary in many characters, but also that local populations differ from each other in the mean values of these characters. In fact detailed studies, such as those on the *Achatinella* snails and the *Peromyscus* mice, had shown that no two populations within a species are ever exactly the same.

Faced with this situation, field workers had come to appreciate the importance of the local population for evolution. The geographical variation from one locality to another evidently formed the link between individual differences and the differences that exist between species. The importance of geographical races was recognized by classifying them as subspecies, through the use of a third name in the Linnaean nomenclature. In well-worked groups of animals, such as birds, butterflies and mammals, it had turned out that the majority of species are made up of local races or subspecies. It was the exception rather than the rule for a species to be uniform throughout its range.

Where a subspecies was quite isolated, it was often a moot point as to whether it should be classified as a subspecies or ranked as a full species. As a result, it was sometimes difficult to be sure where one species ended and the next began as one moved across a large tract of land or a group of islands. By contrast, the distinctions between species at any one locality were nearly always clear cut.

Mayr argued that all this evidence supported the view that new species usually arise from geographically isolated races, which are incipient species. A new species will be formed if characters that reduce or prevent interbreeding with the parental species evolve during this period of isolation. These isolating mechanisms will then promote or ensure 'reproductive isolation' when the geographical barriers break down and contact is made with the parental species once more. One species will then have evolved into two by this process, which Mayr dubbed 'geographic speciation'. So Mayr agreed with Dobzhansky that new species are formed through a gradual sequence of stages, involving a period of geographical isolation for one or more populations of the parental species.

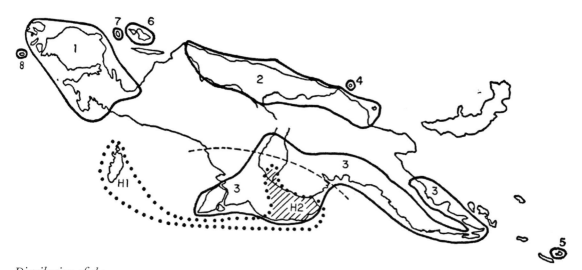

Distribution of the racquet-tailed kingfishers, Tanysiptera galatea *and* T. hydrocharis *in the New Guinea region:* 1 to 3 are the mainland races and 4 to 8 the island races of T. galatea. T. hydrocharis *is present both on the Aru Islands (H1) and on New Guinea (H2). The broken line shows where an arm of the sea formerly existed. From Mayr's* Animal Species and Evolution *(originally published in* Evolution as a Process, *see* Illustration Sources page 276). *A drawing of* T. galatea *can be seen in the lower part of the illustration on page 135.*

Of course, one could not normally expect to see a species go through all these stages in the space of a few years, but one could find species in nature that represented the different stages.

For instance, one case cited from Mayr's own field work was that of the New Guinea kingfishers in the genus *Tanysiptera*. On the mainland of New Guinea, there are three subspecies of *Tanysiptera galatea* that differ from each other only a little. There are also five other subspecies, each of which is confined to one of the smaller islands around the mainland. Each of these forms is quite distinct, and it is difficult to tell whether they are really subspecies of *T. galatea* or are separate species. In addition, there is a form of *Tanysiptera* on the Aru Islands that is also found in South New Guinea, where it lives side by side with *T. galatea* without interbreeding. Hence it is definitely a separate species and is named *T. hydrocharis*.

How this situation came about can be understood in terms of the geological history of the area. The Aru Islands were formerly joined to South New Guinea, making a large island that was separated from the New Guinea mainland by an arm of the sea. The form *hydrocharis* that lived on this island was then one of the isolated island forms that were evidently derived from *T. galatea* on the mainland. But when erosion filled in this arm of the sea, *T. galatea* was able to overlap the range of *hydrocharis* but without interbreeding, showing that *hydrocharis* had become a separate species. There were many such examples that

Mayr was able to cite in discussing species formation, and still more were analysed in the 1940s and 1950s.

Some of the best examples concerned species of birds on groups of islands since even narrow stretches of water are effective, though not insuperable, barriers to land-dwelling birds. One of the most notable studies was the analysis of finches on the Galápagos Islands carried out by David Lack (1910–73). This group of birds became known as Darwin's finches because Darwin had been the first to collect and study them while on his visit to the Galápagos (see colour plate 9). Darwin had been unable to make effective use of the finches in developing his theories as his collections were inadequate, but this had been corrected by the time Lack came to study them. As well as looking at museum collections, Lack was able to visit the Galápagos and make a meticulous study of the living birds.

He decided that Darwin's finches belong to 14 separate species, which differ from each other in the beak, in body size and in male and female plumage. Yet these species are so alike in appearance and behaviour that there could be little doubt that they were all descended from a single ancestral species. Most of these species are found on more than one island of the Galápagos, with the island forms of the same species differing from each other to varying extents. The differences between these island forms involve the same characters — beak, body size and plumage — that distinguish the species. Hence Lack concluded that 'new species have arisen when well-differentiated island forms have later met in the same region and kept distinct'.

The original colonists could have evolved from one into 14 species by colonizing one island after another, and after reaching species level recolonizing islands from which they had come. One instance in which this process of geographic speciation could be followed clearly was that of the two species of large insectivorous tree-finch, *Camarhynchus psittacula* and *C. pauper*. These occur together on Charles Island (now Floreana) without interbreeding, and are undoubtedly two separate species even though the differences between them are small. *C. pauper* is confined to Charles but *C. psittacula* is divided into three well-marked subspecies: *C. p. psittacula* on Charles and on islands to the north, *C. p. habeli* still further north and *C. p. affinis* to the west.

Of these subspecies, the form *affinis* closely resembles *C. pauper* in both beak and plumage, but it also grades into form *psittacula* through a

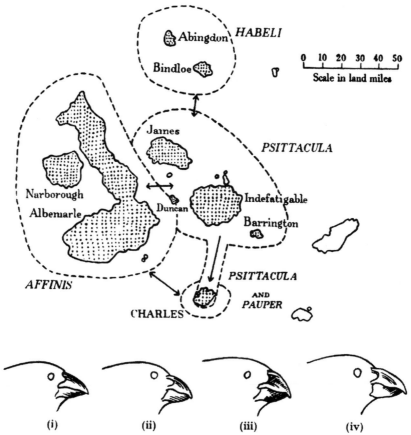

A case of geographic speciation in Darwin's finches on the Galápagos Islands. This map, from Lack's 1947 book, shows the distribution of three subspecies of Camarhynchus psittacula: *forms affinis, habeli and* psittacula. *The form* pauper *would be regarded as a fourth subspecies but for the fact that it overlaps with* psittacula *on Charles (Floreana) without interbreeding. The arrows indicate probable routes of migration between the subspecies.*

population of intermediate type on the small island of Duncan. If the form *psittacula* did not occur on Charles, *pauper* would be regarded simply as a fourth subspecies of *C. psittacula*. As it is, the facts suggest that Charles has been colonized successively by two separate subspecies of the large insectivorous tree-finch. The island was probably reached first by ancestors of the form *pauper* and was later invaded from the north by the form *psittacula*. Originally, as Lack explained, *pauper* and *psittacula* 'were geographical races of the same species, but by the time they met on Charles they had become so different that they did not interbreed, and so they have become separate species'.

In his book on *Darwin's Finches*, published in 1947, Lack did more than give another example of species formation in birds. He also provided good evidence that the main beak differences between the species are adaptations to differences in diet. This was an important point because it was then widely believed that many of the structural differences between species are non-adaptive. The features by which systematists separated one species from another often seemed so trivial that people were reluctant to believe that they were important in the life of the animal. So the detailed differences between species were often viewed as the result of chance effects such as genetic drift due to the isolation of the incipient species. Lack himself had accepted such a view in his first published account of evolution in Darwin's finches.

He changed his mind after realizing that two species with truly similar ecology cannot live in the same region indefinitely. It is a simple consequence of natural selection that two similar species will compete with each other, and the one that does better will eventually eliminate the other. When he looked at his data again, Lack saw that beak structure was clearly correlated with the method of feeding in Darwin's finches. This was shown by the heavy finch-like beak of seed-eating species, the long beak of flower-probing species, the rather parrot-like beak of vegetarian species, and so on. Even in closely related species, where the beaks differed only in size, there was some evidence of a correlation with diet.

Lack concluded that natural selection had played a major role in shaping the beaks of Darwin's finches. Differences between species that had earlier been treated as incidental or due to chance could now be seen as the result of adaptation and competition. Strong evidence that the differences between races or subspecies were also adaptive was provided by another study published in the same year as Lack's book. A set of experiments had been carried out on the light and dark forms of the mouse, *Peromyscus maniculatus*, by L. R. Dice, who built on the earlier work of Sumner.

These experiments were designed to test whether or not the differences in coat colour were adaptive. Individuals of both forms were put into cages with a floor of either dark or light soil and were preyed upon by two species of owl. In all cases, even under the dimmest illumination, those mice which were less conspicuous against their background proved to have a significant advantage in escaping capture. Clearly, selection of this kind could have been responsible for the situation where lighter populations of *Peromyscus* are associated with lighter habitats, and vice versa.

Meanwhile, there was the question of how well these conclusions fitted in with the conclusions that palaeontologists were drawing from the fossil record about evolution on the grand scale. The man who did more than anyone else to integrate the study of evolution at these different levels was George Simpson (1902–84), whose speciality was vertebrate palaeontology. While based at the American Museum of Natural History in New York during the 1930s and 1940s, he took a keen interest in contemporary work on natural selection and species formation. In 1944, he published *Tempo and Mode in Evolution*, which set out to assess the fossil record in the light of the new population studies.

As he warned readers in the introduction, Simpson's book departed from the usual pattern of describing the anatomy of fossils and inferring evolutionary relationships. Instead he took the innovative step of using particular groups of animals, such as carnivores or horses, to try and estimate rates of evolutionary change. He found that the rate of change was far from constant. Within a particular group, a period of rapid change was often followed by a period of little change, and the rate of change also differed widely between one group of animals and another. He also found similar variability in the length of time that particular groups survived in the fossil record.

Simpson then turned his attention to the new genetical work, noting the theoretical studies of Fisher, Haldane and Wright, as well as the studies on natural populations by Dobzhansky and others. He argued that the accumulation of small mutations through natural selection was the mechanism of evolution best supported by the evidence, and one that was able to produce rapid evolutionary change. In his view, no additional forces were required by the fossil data. The patterns found in the fossil record could be understood in terms of mutation and selection at the population level.

In the famous horse series, for instance, evolution did not unfold in a straight line from *Eohippus* to the modern horse in the way that Cope had supposed. Those simplified charts of horse evolution were not accurate, as W. D. Matthew had realized when he complained of a tendency 'to put the chart before the horse'. There was now enough fossil material, Simpson argued, to show that there had been continuous but branching evolution in the horse family. At each main stage, the horse family had given rise to four or five separate lineages, each with its own combination of teeth and toes. The evolution of horses had proceeded in different

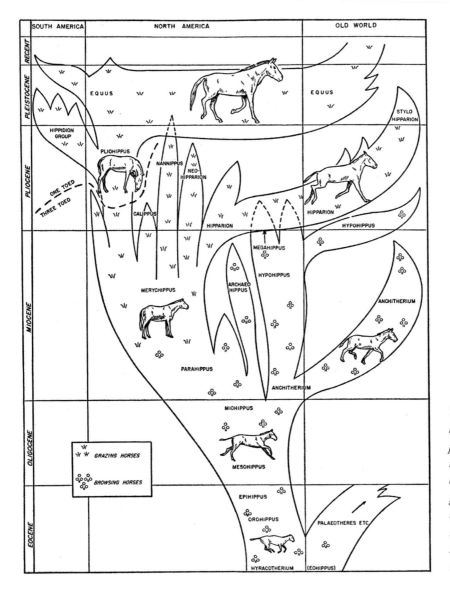

SOUTH AMERICA	NORTH AMERICA	OLD WORLD

The evolution of the horses, a diagram published by Simpson in 1951. The diversity of fossil horses was so great that their evolution could not be interpreted as a straight line leading to the modern horse.

directions and at different rates at different times, as one would expect if variation and selection were the foundation of long-term evolution.

By incorporating the results of genetics and population studies in this way, Simpson's book did much to reduce the gap between palaeontology and these other approaches to studying evolution. So when a group of palaeontologists, geneticists and naturalists met together in 1947 to discuss common problems of evolution, a broad consensus emerged. It was agreed that evolution proceeds gradually and not by sudden jumps, with natural selection being the principal driving force of evolutionary change. Events at the population level were recognized as the key to the formation of new species and hence to the origin of diversity in general. The published account of these discussions, in a book edited by G. L. Jepsen, Mayr and Simpson, was given the bold title *Genetics, Paleontology and Evolution*.

This new agreement among diverse specialists on a number of basic points became known as 'the modern synthesis'. In using this phrase, people did not mean that a common mind had been reached on every topic relating to evolution. Different research groups naturally retained different perspectives, but it was recognized that a substantial area of common ground had been established. Such a synthesis was welcomed as a breath of fresh air after the apparently irreconcilable differences of opinion that had existed less than twenty years before. For it was soon appreciated that a basis had now been laid, on which future evolutionary studies could be safely built.

CHAPTER NINE

THE CONTINUING JOURNEY

'The theory of evolution is not just getting older, it is getting better.' This statement, made by Steven Stanley at the beginning of a book in 1981, nicely reflects the strength of a good theory. If it is to stand the test of time, a theory must be capable of being improved as new evidence is brought in. This is well illustrated by the modern synthesis, in which the new evidence led to an improved version of Darwin's theory and not to its rejection. After the doubts of the early years of the century, the new insights of the 1930s and 1940s confirmed and strengthened the theory of evolution by natural selection. Not that the synthesis was a complete and finished product in itself. Rather, it provided a basis for still further improvements, which took place in the second half of the twentieth century.

By a happy coincidence, the modern synthesis of evolutionary theory arrived just in time for the expansion of the universities in the post-war period. Science in particular was well funded and biological research grew significantly during the 1960s and 1970s. This favourable environment resulted in what might be called an adaptive radiation of evolutionary studies. The growth and variety of research on evolution was such that our narrative cannot follow every line of investigation. Instead it will take a path that picks out just a few of the highlights in this expanding field of research.

Without a doubt, the most significant development in biology during the second half of the twentieth century was the finding that genes are composed of nucleic acid, carried in the chromosomes of each cell. The growth of genetics in the earlier part of the century had led to the identification of more and more genes in well-studied organisms such as *Drosophila*. Eventually, the sheer number of genes located on each chromosome had indicated that each gene must be so small as to be of molecular dimensions. Since chromosomes were known to contain nucleic

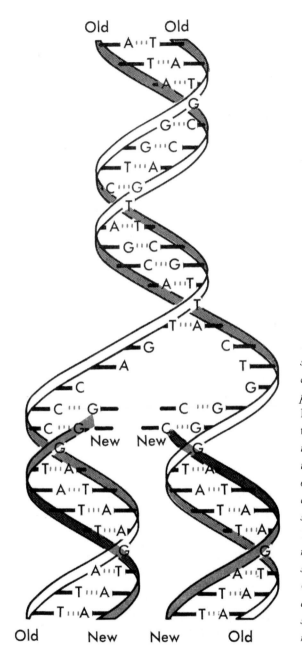

Watson's diagram showing replication of a DNA molecule, published in his Molecular Biology of the Gene *in 1965. As the double helix unwinds, new strands of DNA are formed alongside the separated old strands. The strict bonding rules ensure that the sequence of bases (A, C, G, T) in each new double helix stays the same as in the parent molecule.*

acid and protein, both of these molecules had been canvassed as possible candidates for the genetic material. The answer was provided by James Watson and Francis Crick in 1953, when they elucidated the now-famous double helix structure of deoxyribonucleic acid, or DNA for short.

Their paper showed that the DNA molecule is a long strand made up of repeated units, called 'nucleotides', each of which consists of a phosphate, a sugar and a 'base'. The phosphate and sugar are attached to each other and are also linked to neighbouring nucleotides, so forming the backbone of the DNA molecule. The base of each nucleotide, which is attached to the sugar, sticks out at an angle from this backbone. Most of the time, DNA consists of two such strands, which spiral around each other in a double helix arrangement. The bases of each strand project into the centre of the helix, where they pair up with the bases of the other, 'complementary' strand.

There are four types of base and, when they pair up with the complementary strand, each base bonds strictly with just one of the other types. This arrangement underlies the ability of DNA to carry the genetic information, as Watson and Crick quietly noted. The specificity of base pairing ensures that the sequence of bases along the DNA molecule is preserved when the DNA is replicating during cell division. The two strands of the double helix come apart as the cell divides and each strand can then act as a template for the synthesis of a new complementary strand. In this way, the new double helix of DNA in each daughter cell will have the same sequence of base pairs as the parent cell.

This exciting discovery was part of a whole new approach to biology, which soon became known as molecular biology. One of the first tasks to be undertaken in this new field was to find out how the genetic information is read out from the DNA molecule during development. It was already evident that most components of a cell, and hence of an entire body, are proteins ranging from structural components to enzymes. And research on the oxygen-carrying proteins, haemoglobin and myoglobin, was showing how each protein molecule is made up of a specific sequence of amino acids.

Further research in this area then found that the sequence of amino acids in a protein is specified by the sequence of bases in a given region of DNA. The precise sequence of bases specifying each amino acid was soon worked out and this became known as the 'genetic code'. So an organism could now be seen as a complex system, the development of which is controlled by the genetic information carried in its DNA molecules. This basic insight had been reached in just over a decade, and Watson was able to review progress in his book, *The Molecular Biology of the Gene*, in 1965.

These advances in molecular biology served to strengthen the view of evolution embodied in the modern synthesis. For the discovery of DNA as the genetic material of living things, and the subsequent unravelling of the genetic code, made it quite clear how variation is generated for natural selection to work upon. Mutations, for example, were now seen to be copying errors that occur from time to time when the DNA is replicating within a cell. Any such error in copying the sequence of bases could alter the genetic information they carry, much as a typographical error can alter the meaning of a word or sentence. The effect of that alteration in a germ cell becomes apparent when the genetic information is read out during development to give a particular bodily structure. The resulting organism then has to make a living in its natural environment, which is where natural selection comes into play.

Beyond this, the molecular biology of inheritance opened up fresh avenues of enquiry, including a new approach to the study of evolutionary relationships. Previously the evolutionary relationships among groups of organisms had been inferred mainly on the basis of differences in bodily form and development. It was now clear that these differences result from changes in the base sequence of DNA molecules. Therefore, a comparison of base sequences in the DNA of different groups ought to yield a reliable measure of how closely they are related.

At first, it was not possible to determine the sequence of bases in a DNA molecule. Consequently a variety of less direct methods were used to compare the DNA in a range of living organisms. The simplest comparison, which was used early on, is to estimate the proportion of bases that differ between two species, and this measure became known as the 'genetic distance'. Such a measure could be obtained at second hand by comparing the similarity of a specific protein in different species, and the technology to do this was available by the 1960s.

The new methods were applied to a variety of animal groups but the comparison between humans and other primates naturally excited the most interest. That humans are closely related to the great apes had been recognized ever since Huxley had published *Evidence to Man's Place in Nature* in 1863, but many uncertainties remained. Anatomical comparisons had not been able to say which species of great ape is our closest relative, although the gorilla and chimpanzee of Africa were generally preferred to the orang-utan of Asia. Two key papers in applying molecular studies to the ape/human comparison were published in 1966 and 1967 by V. M. Sarich and A. C. Wilson, who used immunological methods to assess the similarity of protein molecules.

Sarich and Wilson began by generating antibodies to a single protein from human blood and then tested the reaction of these antibodies to the corresponding protein from several species of ape and monkey. They found that the level of reaction produced by both chimpanzee and gorilla protein was 95% of that given by the human protein, while that of the orang-utan was only 85%. Similarly, the reaction produced by a gibbon (or lesser ape) from Asia was 82% and that of an Old World monkey (one from Africa or Asia) was 73%. The differing reaction levels reflected the degree of difference in the proteins, which in turn must reflect the difference in base sequences of their DNA. In the case of humans and the African apes, this genetic distance of 5% was remarkably small, comparable to that between a horse and zebra, for example.

The most innovative feature of Sarich and Wilson's work was that they used the molecular comparisons to estimate how long ago the human lineage had split off from the great apes. The genetic distance between two species of primate should be proportional to the time since they separated in evolution provided the rate of change at the molecular level had not varied greatly. Citing evidence in support of this assumption, Sarich and Wilson calibrated their estimate by using a date of 30 million years ago for the split between Old World monkeys and the ape/human lineage. Their calculations then yielded a date of about 5 million years ago for the split between African apes and humans. This took many people by surprise since a date of 15−20 million years ago was widely thought to be about right at that time.

A few years later, it became possible to measure the genetic distance between two species by means of DNA hybridization. This method compares two entire DNA molecules extracted from the cells of different species. The two strands of each molecule are separated by heating and then the single strands from two different species are put together so they can join up. This produces a hybrid DNA molecule, which can in turn be separated by heating. The stability of this hybrid molecule, and hence the heat required to separate its two strands, depends on how similar the base sequences are in the two species. So the temperature at which separation takes place can be used to measure the overall similarity of DNA in pairs of species.

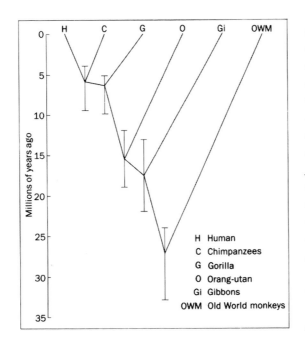

The relationship between living apes and humans based on a consensus of DNA comparisons, available by the end of the 1980s. The diagram also shows the estimates, derived from these DNA comparisons, for the times of separation between the different species and or higher groups. From the Cambridge Encyclopedia of Human Evolution, *published in 1992.*

This method was applied to the DNA of apes and humans by C. G. Sibley and J. E. Ahlquist in 1984. They found a smaller genetic distance between chimpanzees and humans than between either of these species and the gorilla. This finding supported the view that the chimpanzee, rather than the gorilla, is our closest living relative. In fact, their work produced the much-quoted result that 98.5% of human DNA is identical to that of chimpanzees. When this information was used to estimate a date for the split between chimpanzee and human lineages, a figure of between 5 and 6 million years ago was obtained.

During the 1980s it became possible to determine the sequence of bases in a particular region of DNA using direct methods. As the techniques for DNA sequencing became available, they were used by a number of researchers to compare a given gene in different species. The corresponding gene could be sequenced in several species and the resulting sequences lined up alongside each other to identify the differences. One of the first DNA regions to be compared across different species of primate was the family of globin genes that code for parts of the haemoglobin and myoglobin molecules.

These studies on primates yielded results that fitted in well with the conclusions drawn from the comparison of proteins and from DNA hybridization. The comparisons of DNA sequences confirmed that genetically the African apes are closer to humans than the orang-utan from Asia and that chimpanzees are our closest living relative. A consensus thus emerged from the various molecular studies concerning the evolutionary relationships between living apes and humans. This consensus brought an important clarification to what had previously been an area of uncertainty, and it has held up well in the light of many later studies. This is just one among many examples of the way in which molecular biology has contributed to the study of evolution.

At the same time, traditional areas of evolutionary study continued to flourish and even expand in the second half of the twentieth century. One such area is the study of fossils, which was helped along by exciting new finds, including a number of significant vertebrate fossils. Fossil discoveries in the early part of the twentieth century had made it possible to trace the course of vertebrate evolution with increasing confidence, as we saw in chapter 7. Fossils found in the second half of the twentieth century not only provided a more detailed picture but also helped to illuminate major points of transition such as that from reptiles to mammals and from primate ancestor to the human species.

In the case of the reptile to mammal transition, the earlier work had brought to light a succession of reptiles through the Permian and early Triassic periods that became more and more like mammals. This trend is seen in the lower jaw, where the bones other than the dentary became progressively reduced to tiny remnants at the back of the jaw. However, the mammal-like reptiles of the early Triassic still retained a reptilian jaw joint between the reduced articular bone in the lower jaw and the quadrate bone of the skull, whereas in mammals the dentary is the only bone in the lower jaw and it contacts the squamosal bone at the back of the skull (see illustration on page 180).

Some of the newer fossil finds showed how the transition from the reptilian jaw joint to the full mammalian condition was accomplished. A lower jaw and skull fragments from the late Triassic were found in the 1960s, and named *Diarthrognathus*. These animals still had a jaw joint composed of the reduced articular and quadrate bones but the dentary also had a backward extension or condyle, which just made contact with the squamosal on the skull. The same is true of *Probainognathus*, a more complete specimen of a mammal-like reptile from the mid Triassic, which was described by A. S. Romer in 1970. Here too there was one jaw joint composed of

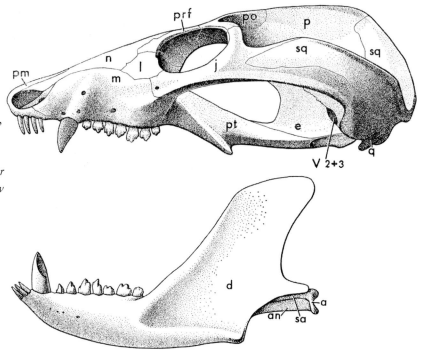

Skull and lower jaw of the mammal-like reptile, Probainognathus, *described by Romer in 1970. The small articular bone (a) of the lower jaw contacts the quadrate bone (q) on the skull. In addition, the dentary (d), which makes up most of the lower jaw, extends back alongside the articular to make contact with the squamosal (sq).*

the articular and quadrate and another composed of the dentary and squamosal bones as in mammals. In effect, a reptilian and a mammalian jaw joint were present side by side. So these fossils mark a clear transition to the earliest known mammals, which appear at the end of the Triassic.

Fascinating though these fossils are, in terms of popular appeal they were no match for the new finds relevant to human origins, which were coming to light at roughly the same period. The situation at the start of this period can be illustrated appropriately by a paper published in 1950 by W. E. Le Gros Clark at Oxford University. In this paper, Le Gros Clark made a detailed analysis of the teeth of apes and humans, carefully noting the differences between them. Then he compared these with the fossil teeth and jaws of a possible candidate for human ancestry named *Australopithecus africanus*. This name had been given to a fossil juvenile skull described by R. Dart in 1925.

Further finds of adult skulls were made in 1936 and 1947 by Broom, who had also studied the mammal-like reptiles (he gave the skulls a different name, later changed to *Australopithecus*). The *Australopithecus* fossils were significant on account of being the oldest human fossils known at that time. Their skulls were more ape-like than human in overall

217

appearance, with long jaws jutting out from the face and a relatively small cranium to accommodate the brain. Nevertheless, Le Gros Clark's study showed that they were consistently closer to humans than apes in the details of their teeth. He therefore supported the inclusion of *Australopithecus* in the human family, the Hominids.

This view was also supported by the few bones that had been recovered from the rest of the *Australopithecus* skeleton, in particular an almost complete pelvis and parts of the femur (thigh bone). These fossil bones resembled the corresponding bones in humans much more than those in apes, which suggests that their owners also resembled humans in walking upright as habitual bipeds. An upright stance was further suggested by a study of the occipital condyles, which connect the back of the skull to the top of the vertebral column. The placement of these condyles in the *Australopithecus* skulls resembled the human pattern rather than that in apes. And so it came to be generally accepted that these were early hominids, which had walked upright but had relatively small brains.

Another noteworthy paper published in 1950 was written by Ernst Mayr, who used his understanding of biological species to sort out the classification of fossil hominids. Hitherto, a new find that was even slightly different from previous ones often got named as a new species or even a new genus of hominids. Mayr attempted to restore some biological order to this situation by reducing the number of species and genera. In some respects he overdid this, but one name that he proposed has stood the test of time.

For among the fossils he reviewed were those popularly known as 'Java man' and 'Peking man', which had been found on the island of Java and in a cave near Peking (now Beijing) in China. These locations had yielded many specimens in the previous 15 years, including several well-preserved skulls and a reasonable sample of bones from the rest of the skeleton. Earlier workers had assigned Java man and Peking man to separate species and given them their own genus, *Pithecanthropus*. But Mayr concluded that they were sufficiently similar to each other to be placed in the same species and sufficiently similar to modern humans to share our genus. Accordingly, he put them together in the species *Homo erectus*, and this was soon accepted as a sound classification.

The inclusion of this species in the genus *Homo* made sense because, from the neck downwards, the skeleton was almost identical to that of a well-built modern human. The skull, however, differed from that of a

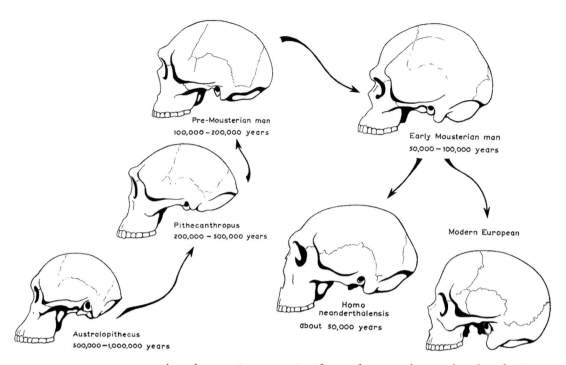

Pre-Mousterian man
100,000 – 200,000 years

Early Mousterian man
50,000 – 100,000 years

Pithecanthropus
200,000 – 500,000 years

Modern European

Australopithecus
500,000 – 1,000,000 years

Homo
neanderthalensis
about 50,000 years

The sequence of human evolution according to Le Gros Clark, based on the fossil skulls known in 1959.
Pithecanthropus *would now be called* Homo erectus, *Early Mousterian man would be* Homo heidelbergensis *and Modern European is* Homo sapiens *(Pre-Mousterian man is a single fossil with intermediate features). Note that the only branching point in this sequence is where* Homo neanderthalensis *separates from the main line of human evolution.*

modern human in some significant features, hence the classification as a separate species. The face had a large jaw with distinctive teeth and also had pronounced eyebrow ridges. Above these brow ridges, the skull sloped back at a shallow angle, enclosing a brain of modest size. The size of the brain could be inferred from the volume of the space inside the skull, which was between 900 and 1000 ml in these specimens. This is about two-thirds of the size of an average modern human and about twice that of *Australopithecus africanus*.

From the structure of its skeleton, *Homo erectus* was well qualified as an intermediate between *Australopithecus africanus* and *Homo sapiens*. It was also intermediate in time, judging by the deposits in which fossils of the three species had been found. This seemed to make a straightforward story of evolutionary transition from *Australopithecus* through *Homo erectus* to early *Homo sapiens*, with the neanderthals as a fairly recent side branch. Such a straightforward picture was accepted, with due caution, by Le Gros Clark when he published a book entitled *The Fossil Evidence for Human Evolution* in 1964. But a paper published by L. S. B. Leakey and colleagues in the journal *Nature* in that same year began to complicate the story.

Louis Leakey was born in Kenya of English parents and soon developed an enthusiasm for African prehistory. Early in his career, he saw the

potential of the African Rift Valley as a location for finding fossil hominids. At his favourite site, Olduvai Gorge in Tanzania, he discovered
stone implements of a crude design, which he named Oldowan stone
tools. His search for early hominids, who could have made the tools,
was rewarded with a remarkably complete skull of *Australopithecus*,
found by his wife Mary in 1959. This find was a robust form of
Australopithecus, with huge cheek teeth, similar to some specimens
found in South Africa by Broom. It was widely felt that these represented
a separate species from *Australopithecus africanus* and this impression was
reinforced by the new find.

But people were reluctant to accept the relatively small brained
Australopithecus as a maker and user of stone tools, however basic. So it
was a delight for Leakey and his colleagues to be able to announce
their find of a larger-brained hominid at Olduvai in the 1964 paper.
Although the four skulls found were fragmentary, it proved possible to
estimate the size of the brain cavity at around 640 ml, which was significantly larger than the values for *Australopithecus*. This was enough to convince the authors not only that these specimens represented the true
toolmakers, but also that they belonged in the genus *Homo*. So their
paper assigned these finds to a new species called *Homo habilis*, meaning
'handy man'.

Due to the incomplete nature of the material, a good deal of debate took
place as to whether the new name was justified. Over the years it was gradually agreed that these fossils were close to the borderline between
Australopithecus and *Homo*. What was even more significant was the fact
that some of the *Homo habilis* specimens had been found in the same
deposits as the robust *Australopithecus*. This meant that, contrary to the
simple picture of human evolution, a larger-brained species of hominid
had existed alongside the smaller-brained species at this early stage.

It was possible to say just how early this was because the
potassium—argon dating method had recently become available. This
method is suitable for dating rocks that are millions of years old and is
most reliable when used on volcanic rocks such as lava flows or tuffs.
When applied to the rocks of Olduvai Gorge, the potassium—argon
method showed that the robust *Australopithecus* and *Homo habilis* both
dated back to 1.75 million years ago. This estimate produced quite a
shock at the time because it was so much older than people had imagined
for the these rocks, which belonged to the Pleistocene geological period.

Over the following decade, a number of research groups took the search for hominid fossils further north along the African Rift Valley. One important site was the shore of Lake Turkana in Kenya, where the Leakys' son, Richard, set up a team of fossil hunters. Their ongoing work resulted in a series of significant finds, beginning with some skulls identified as male and female specimens of the robust *Australopithecus*. Then in the mid 1970s, an exceptionally complete skull was found and dated to between 1.3 and 1.6 million years ago. This skull, known by its museum number ER 3733, was so similar to the hominid fossils from China that it was assigned to the species *Homo erectus*. Earlier fragmentary finds had raised the possibility that *Homo erectus* had lived in Africa as well as Asia but ER 3733 placed the matter beyond doubt.

Meanwhile, a joint French and American team had begun working at the northern end of the African Rift Valley, in the Afar region of Ethiopia. Their endeavours led to the truly remarkable find, made by Donald Johanson late in 1974, of a relatively complete hominid skeleton. This skeleton was given the nickname 'Lucy' and justly acquired celebrity status on account of its great age as well as its completeness. The rocks from which Lucy was extracted belonged to the Pliocene period, and their age was carefully dated to 3.2 million years ago, making this by far the oldest hominid fossil found so far. And as some 40% of the skeleton was represented, it was possible to reconstruct Lucy's appearance accurately, with little room for fanciful interpretation (see colour plate 16A).

Accurate reconstruction was also helped by more finds at the same locality, including the fragmentary remains of several individuals, similar in structure to Lucy. A detailed appraisal of all this material made it quite certain that these early hominids were bipeds, walking upright as we do. Skeletal parts such as the pelvis, knee joint and foot bones were so similar to those of modern humans that their mode of locomotion must have been essentially the same. However, there were subtle differences in the relative width of the pelvis, the length of the legs and curvature of the toes, which suggested that their bipedalism may not have been as refined as that of modern humans.

But although these hominids were bipeds, they were not tall. Lucy's height can be estimated from the relatively complete skeleton and it comes out at not much over 1 m tall. Dramatic confirmation of both bipedalism and height came in 1978, in the form of fossilized footprints uncovered by Mary Leakey's team at Laetoli in Tanzania. The footprints had been made in an ashfall from a nearby volcano, enabling them to be reliably dated by the potassium—argon method at 3.7 million years old. The wonderful preservation of these ancient footprints clearly showed the features characteristic of bipedal walking, such as a well-developed heel, a strong arch and the big toe pointing straight forward.

The skull of Lucy and her kind was the least complete part of their fossil skeletons, but enough fragments were found to make a good reconstruction. The general appearance was rather ape-like, similar to the skull of *Australopithecus africanus*. But the details of the jaws and teeth, while still intermediate, resembled those of the African apes more than those of humans. The cranium was definitely small and estimates from four specimens yielded a value of about 440 ml for the brain space, which is only slightly larger than that of the average chimpanzee. Thus the picture of early hominids as bipeds with small brains and protruding jaws was strongly confirmed by this new material.

Where then did Lucy and associated finds fit in to the scheme of classification? Johanson and his colleagues had not named their finds at first but waited until the fossils had been fully described. They then assigned the name *Australopithecus afarensis*, after the Afar region in which the fossils had been found. The excellent material assigned to *Australopithecus afarensis*, and the associated footprints at Laetoli, helped to make the 1970s a significant milestone in research on hominid fossils. But discovery and analysis showed no sign of slackening off in the 1980s.

The find of the decade was made by Richard Leakey's team in 1984, to the west of Lake Turkana. There they discovered a hominid skeleton which was even more complete than that of Lucy. This skeleton belonged to a young male individual and so acquired the nickname 'Turkana Boy', and it was dated to about 1.6 million years ago (see colour plate 16B). The discoverers of Turkana Boy had no hesitation in placing this find in the species *Homo erectus*, given the details of its skull and the rest of the skeleton.

This remarkably complete skeleton showed that the limb bones were almost identical to those of modern *Homo sapiens* in both their size and proportions. It was estimated that, as an adult, Turkana Boy would have grown to 1.8 m (6 feet), fully as tall as modern humans. The well-preserved skull had the prominent brow ridges and low forehead so characteristic of *Homo erectus* and its brain space was about 850 ml, much the same as skull ER 3733. This value was larger than that of earlier hominids but still quite small compared to *Homo sapiens*. So the skeleton was like that of modern humans from the neck down but this important transition was accompanied by only a modest increase in brain size.

While these fossils of *Homo erectus* in Africa could be regarded as intermediate between earlier hominids and *Homo sapiens*, there were other hominid fossils for which this obviously was not the case. In 1985, deposits west of Lake Turkana yielded the oldest skull yet of a robust *Australopithecus*, dated at 2.5 million years old. This skull was sufficiently different from the other robust *Australopithecus* found in East Africa for it to be classified as a separate species. In fact, discussion around this time led to the suggestion that these robust species should be ranked in a separate genus. And the generic name *Paranthropus*, originally proposed by Broom for the South African species, was put forward. Here, then, was a distinct group of fossil hominids that were clearly not on the line of evolution leading to *Homo sapiens*.

The level of species diversity was also increased by further finds in the Olduvai Gorge region, which led to the conclusion that the material classified as *Homo habilis* might in fact represent more than one species. Then in the 1990s, work at sites in Ethiopia and in East Africa uncovered hominid fossils that dated back more than 4 million years and were evidently distinct from *Australopithecus afarensis*. Overall, the range of fossil hominids known by the end of the twentieth century could no longer be viewed as a simple sequence leading to *Homo sapiens*, but rather resembled the pattern of adaptive diversity seen in other groups of animals.

Another field of evolutionary study that blossomed over this period was research on natural selection as the main mechanism of evolution. The power of selection to change the hereditary make-up of a population had been demonstrated by mathematical studies and by breeding experiments earlier in the twentieth century, as discussed in chapter 8. Subsequently, a range of field work began to study natural selection as a process taking place in natural populations. By 1986, a book on *Natural Selection in the Wild* by John Endler listed well over a hundred species, in which natural selection had been directly demonstrated in field studies.

One of the most detailed of these studies was initiated by Peter and Rosemary Grant in 1973, when they made their first visit to the

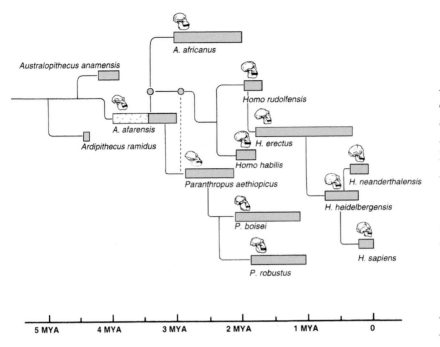

Diagram of hominid evolution from a textbook on Primate Adaptation and Evolution *by John Fleagle, published in 1999. The shaded bars indicate the known duration of each species relative to the timescale below. Note the large number of species and branching points in this phylogeny.*

Galápagos islands. Their purpose was to take a fresh look at Darwin's finches. Whereas Lack and others had been able to infer the role of natural selection in shaping Darwin's finches, the Grants wanted to observe the actual process in a natural population. What made Darwin's finches so suitable for such a project was that they were a well studied group of species living in isolation on remote islands. The Galápagos islands also had the advantage of having remained relatively unspoiled by human activity and the birds there were still unafraid of human visitors.

The Grants chose Daphne Major, a small uninhabited island in the Galápagos, for an intensive study. Here their team of researchers were able to analyse an isolated population of one species of Darwin's finches in great detail. For instance, they could identify every finch on the island and follow the fortunes of individual birds as they fed, mated and raised their offspring. They soon realized the benefit of comparing one year with another, and the initial project grew into a study that has continued on into the twenty-first century.

The species that became the chief focus of this study on Daphne was one of the ground finches, which are classified in the genus *Geospiza*. As their name implies, these species of Darwin's finches feed on or close to the ground, which makes for convenient observation, and three of them have beaks specialized for cracking and eating seeds.

These three species differ from each other in the mean size of their beaks and so are known appropriately as the small, medium and large ground finches (see colour plate 9). All three species occur on some of the larger islands but only the medium ground finch, *Geospiza fortis*, regularly lives and breeds on Daphne Major.

Detailed observations on Daphne Major showed that the medium ground finches do indeed eat the seeds of plants growing on the island. There proved to be significant variation in beak size within this population, and birds with larger beaks were seen to crack open fairly large and hard seeds that those with smaller beaks could not handle. Their observations showed that this variation was important during the dry season, in the second half of each year, when the finches were dependent on the limited supply of seeds left over from the wet season early in the year.

While this study was under way, it so happened that the wet season of 1977 failed and there was a long dry period on Daphne, from mid 1976 to the end of 1977. During this drought, the numbers of *Geospiza fortis* fell dramatically in keeping with the decline in seeds from the island's plants, on which the finches depend for food. As the number of birds continued to fall, survival did not occur at random: large birds survived better than small ones. This was because the supply of smaller and softer seeds, which are easy for the finches to deal with, was soon exhausted and the birds were forced to tackle larger and harder seeds. So strong natural selection occurred in favour of the birds with larger beaks.

For such an episode of natural selection to cause evolutionary change in the next generation, the features of the beak must be inherited. It is possible to estimate the heritability of certain features in a species by comparing the average measurements of these features in parents and their offspring. Such a comparison was carried out in the population of *Geospiza fortis* on Daphne island, using the researchers' knowledge of individual birds. Measurements of beak and body were made on nesting pairs and the same measurements were subsequently made on their offspring, when they had grown up. The results showed a strong correlation between parents and offspring, indicating that about 80% of variation in beak size and shape is inherited.

So it is not surprising that a significant evolutionary change was seen in the next generation of finches, that is the offspring of those birds that survived the drought of 1976—7. From measurements made on

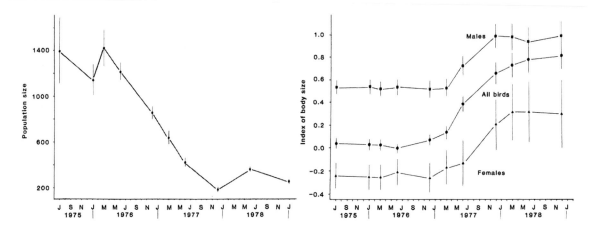

individual birds, the mean values of the population were compared for birds before the drought in 1976 and the next generation in 1978. The comparisons were made on features such as the length of the wing and the width and depth of the beak. The mean value of beak depth, for example, was 9.21 mm in 1976 and 9.70 mm in 1978, which is a 5% increase. This is a remarkably large shift in one generation and hence is an example of rapid evolutionary change in a natural population.

The ongoing study of this population of *Geospiza fortis* saw more evolutionary changes but none was as large as this one in the 1970s. Another significant change was observed in the 1980s but this was in the opposite direction. A major weather change in 1983 caused ten times the normal amount of rain to fall on Daphne and this resulted in an exceptional growth of plants with small seeds. These then dominated the food supply for the finches in the dry years of 1984–6, so this time there was strong selection in favour of smaller beaks. In the case of beak depth, there was a 2% decrease in the next generation.

Overall, this study of Darwin's finches showed how natural selection could be seen in action by maintaining detailed observations of a small, isolated population over many years. It was then possible to witness natural selection as it caused small evolutionary changes from generation to generation, adapting the population to local conditions. The changes observed in *Geospiza fortis* came about because a changing environment, in this case changing weather patterns, forced a shift in diet on the finches, and this exerted a strong selection pressure on the local population.

It is usual to think of natural selection acting on the details of an animal's body such as beak size in birds. But natural selection is quite

Natural selection measured in one of Darwin's finches, Geospiza fortis *on Daphne Island in the Galápagos. LEFT, the decline in the population of* G. fortis *during the drought of 1976–7. RIGHT, the increase in the average size of birds surviving into 1988, as reflected in an index of overall body size. Reproduced from* Science, *with permission (see Illustration Sources page 277).*

capable of producing adaptive changes in any aspects of an animal's life that are under some degree of genetic control. That includes aspects of reproduction, like the age at which individuals become sexually mature and the number of offspring produced at a time. In a given species, the particular combination of these features related to reproduction is known as its 'life-history strategy'. The link between this strategy and the environment in which a species lives has long been a topic of interest to biologists.

A common assumption here is that natural selection will tend to maximize the number of offspring that an individual can produce potentially in its lifetime. And in order to achieve this, the life-history strategy will be modified through evolution to suit the particular circumstances in which a population finds itself. One field study that has put these ideas to the test was carried out by David Reznick and his fellow researchers, starting in the 1970s. They looked at natural populations of guppies, which are small freshwater fish often kept in aquaria and classified in the species *Poecilia reticulata*.

This study was carried out on the island of Trinidad, where the guppies live in rivers that run down the slopes of the mountains. In the lower portions of these streams, the guppies are preyed upon by cichlid fish, which can easily swallow them whole and actually prefer the large guppies. The cichlids are often excluded from the upper portions of these streams by waterfalls or rapids and here the only guppy predators are killifish. These cannot swallow a guppy whole but they prey occasionally on small individuals as part of a more general diet. For the guppies, the risk of predation is naturally much higher where the cichlids are present, and their research found that the mortality rate is indeed significantly higher at these locations.

During this study, guppies were collected from these high-predation sites where cichlids are present and from low-predation sites where there are killifish but not cichlids. All the collected guppies were then kept in aquaria under the same conditions so as to compare their life-history strategies. It was found that the guppies from high-predation localities always reached sexual maturity at an earlier age and smaller size compared to those from low-predation sites. They also had more and smaller offspring each time they gave birth and they gave birth more frequently than the guppies from low-predation sites. Since the guppies were all kept under the same conditions, there was clearly a genetic

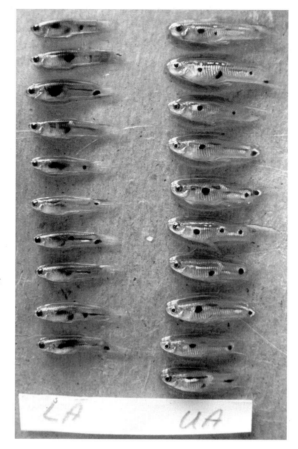

Adult male guppies collected from two parts of a river in Trinidad. Those from the lower part of the river (LEFT COLUMN), where cichlid predators are present, do not grow as large as those from the upper part of the river (RIGHT COLUMN), where cichlid predators are absent.

basis for these differences, and the heritability proved to be higher in males than in females.

It can be shown theoretically that where mortality is high, a good strategy is to produce lots of offspring as early as possible in life. So the contrast between guppies from high-predation and low-predation sites makes sense as an adaptation to the level of mortality at each locality. To test whether this was really the case, Reznick and his team exposed guppies to an altered mortality rate by transplanting them. They found a couple of places where there were cichlids and guppies as usual below a waterfall but above the waterfall there were only killifish and neither cichlids nor guppies were present. Some guppies from below the waterfall were then taken and placed in the pool above the waterfall, so that they experienced a sudden drop in predation level.

The researchers returned at intervals to the locations where the transplants had been made in order to study the descendants of the transplanted guppies. They found that the features of their life history evolved in the direction expected for a low-predation site. These guppies reached sexual maturity at a later age and larger size compared to those still living in the presence of cichlids below the waterfalls. The females produced fewer and larger offspring each time they gave birth, but this change took place more slowly. The most rapid evolution was found in the males, which changed as much as expected in only four years, a rate of evolution comparable to that seen in Darwin's finches.

This outcome confirmed the view that differences in life-history strategies between the populations of guppies have evolved as a direct response to different levels of predation. The features of their life history are evidently adaptations brought about by natural selection, just as much as the structural features of their bodies. This work also shows how much can be learnt about the process of evolution from comparative studies and a simple experiment carried out in a natural setting. Like Darwin's finches, the Trinidad guppies have also proved suitable for a long-term study that has continued on down the years.

Another area in which the study of natural selection has borne fruit during the second half of the twentieth century concerns the behaviour of animals. In particular, fresh insights emerged about forms of social behaviour that seem altruistic. Animals in the wild can often be seen behaving unselfishly towards other members of their own species: sharing food, warning of predators, removing parasites, fighting without injuring their opponents and so on. At first sight, behaviour of this kind seems inconsistent with natural selection, which surely ought to promote an animal's own welfare above all. As Darwin wrote, it 'will never produce in a being anything injurious to itself, for natural selection acts solely by and for the good of each'.

For many years, this problem was not recognized as such because the unselfish behaviour is so obviously of benefit to the species as a whole. The first person to see the problem clearly was William Hamilton, who promptly set about finding an answer to it. The matter was drawn to his attention by warning coloration in insects such as the caterpillars of the cinnebar moth, which flourished near his home in England (see colour plate 12A). The warning colours are effective because naive birds learn to avoid the caterpillars altogether, after first sampling one and promptly rejecting it on account of its unpleasant taste. Although this benefits the moth species, it does not do much for the individual that was sampled first. The mangled creature is hardly in a position to pass its genes on to the next generation. How then could the warning colours have evolved through natural selection?

The key to the problem, Hamilton realized, lay in the fact that the mangled victim and the surviving caterpillars on the same plant were probably close relatives. Very likely they had all hatched from a single batch of eggs. Kinship was the key factor in making sense of social behaviour since natural selection could favour help given to relatives. The germ of this idea had been present in the books of Fisher and Haldane in the 1930s. But it was Hamilton's papers in the early 1960s that explicitly developed 'kin selection' into a general principle, and showed that it was a simple consequence of basic Darwinian theory.

The essential point is that all organisms eventually die, and they pass on not themselves but an internal representation of themselves in their genes. Through the process of reproduction, genes are replicated, shuffled with those of sexual partners and passed on to successive generations. Therefore, what matters in terms of natural selection is success in leaving copies of one's genes to future generations. So an animal that gives help to a relative is increasing the chance that the genes they have in common will survive. The closer the relationship between individuals, the more genes they will have in common and the greater will be the evolutionary benefit of co-operating rather than competing. Hence it does not matter if an animal jeopardizes its own life, provided that in doing so it contributes to the survival of copies of its genes in other animals.

The consequences of this situation were calculated in some detail by Hamilton in the 1960s, and this gave a crucial insight into the evolution of social behaviour. Fortunately, it was relatively straightforward to test these ideas by observing the behaviour of animals in the wild. From the 1970s onwards, numerous field studies were carried out in which close attention was paid to the genetic relationship between the members of animal societies.

One dramatic example is a pride of lions, which consists of two or more males holding a group of females. The males hardly ever fight over access to the females, and they join together to defend the pride against takeover by other groups of males. Field studies have shown that in such a group, the males are usually brothers, and so their cooperation makes sense in terms of Hamilton's ideas. Their tenure of the group of females is often short, and newly installed males may take the drastic step of killing young cubs already in the pride. Consequently, the females soon become receptive to the new males and produce new cubs. By this action, as by their cooperation, the males are behaving in a way that maximizes the contribution of their genes to the next generation.

This is just what one would expect if the behaviour has evolved through natural selection, which alters the frequency of genes from generation to generation. This process can sometimes result in outcomes which we find counter-intuitive, as in this case of newly arrived males killing infants of their own species, which have been fathered by other males. However, this behaviour has been confirmed in other mammals, including primates such as baboons and langurs, on which the most detailed studies have been made. In other respects, too, a whole range of studies have confirmed that natural selection must be important in shaping the evolution of behaviour. This area of research became known as 'behavioural ecology' and grew steadily during the 1970s and 1980s.

For a final example of the expanding research on evolution during the latter part of the twentieth century, let us take another look at the fossil record, only this time the focus will not be on the origin of various groups of organisms, but rather on their extinction. It has been known since the time of Cuvier that extinctions are part of the history of life on Earth, and it was Cuvier who first suggested that extinctions might be caused by large-scale catastrophes. Later in the nineteenth century, Phillips had named three main eras in the history of life, the Palaeozoic, Mesozoic and Cenozoic. His study of the fossil record showed that each of these was separated from the next by a major episode of extinction and renewal.

A century later, Simpson wrote a book reviewing *The Major Features of Evolution*, in which he confirmed that there had been times when the level of extinction reached a peak. And he used the term 'mass extinctions' when referring to the episodes Phillips had recognized between the main eras of life. However, he was not hopeful regarding the possibility of our learning much more about the nature and causes of these extraordinary events. But then, in the early 1980s, two independent studies opened up an exciting new chapter in research on mass extinctions and their possible causes.

The first of these studies dealt with the mass extinction that occurred at the boundary between the Cretaceous and Tertiary periods, which is also the boundary between the Mesozoic and Cenozoic eras. This event has long been famous because it saw the extinction of the dinosaurs on land

and of the ammonites in the sea. A new approach to this extinction came from a father and son team, Luis and Walter Alvarez, who had got together to examine a rock sample from Italy. The sample had three layers: a late Cretaceous limestone and an early Tertiary limestone with a thin layer of clay in between them. Walter, a geologist, appealed to Luis, a physicist, for help in dating the clay layer with the latest methods.

A variety of tests were applied to the clay layer and one of them showed that it had an exceptionally high concentration of the element iridium. Now iridium is rare on the Earth's surface, and what little there is mostly comes from space, arriving in a succession of meteorite showers. Subsequent calculations suggested that the amount of iridium found in the clay layer could have been delivered to Earth by the impact of an asteroid some 10 km in diameter. What is more, the new dating method put the clay layer at 65 million years old, right on the boundary between the Cretaceous and Tertiary. So Luis and Walter Alvarez proposed that the mass extinction at this boundary was actually caused by a large asteroid colliding with the Earth. Their hypothesis was published in *Science* in 1980 and immediately excited a great deal of attention.

People from a range of scientific disciplines moved quickly to check on this idea, and the outcome was favourable for the impact hypothesis. A thin layer of clay was found around the world at the Cretaceous–Tertiary boundary, and it almost always showed high iridium levels. Other indicators that are associated with known impact sites, such as shocked quartz crystals, were also found in the clay layer. Even better, a likely candidate for the crater caused by the asteroid impact was found under the Yucatan Peninsula in the Gulf of Mexico. This find was based on earlier geophysical surveys and revealed a crater of the right age and of a suitable size to have generated the clay layer. And so, by the mid 1990s, there was strong evidence that a large asteroid impact had occurred at the Cretaceous–Tertiary boundary.

That this impact may have finished off the dinosaurs and ammonites, as well as many other groups, was confirmed by a renewed study of fossils from the late Cretaceous. The overall impression gained from these new studies was that some dinosaurs and ammonites became extinct during the late Cretaceous but many continued right to the end of the Cretaceous but not beyond. From this, it seemed most likely that the asteroid impact delivered the final coup for a range of groups, which were already experiencing declines for other reasons. By the end of the

Raup and Sepkoski's graph, published in 1982, showing how the increasing diversity of marine animals over geological time has been interrupted by five mass extinction events. These occurred late in the Ordovician (1), Devonian (2), Permian (3), Triassic (4) and Cretaceous (5) periods. Reproduced from Science, *with permission (see* Illustration Sources *page 277).*

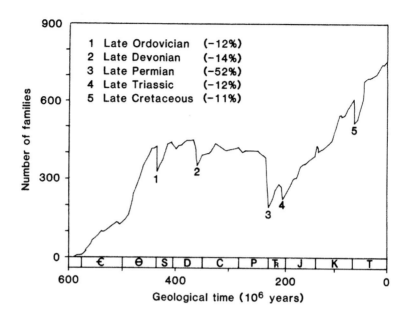

twentieth century, therefore, an asteroid impact was accepted as a genuine cause, but not the only one, for the mass extinction at the end of the Cretaceous.

The second line of enquiry relating to mass extinctions was based on the statistical analysis of patterns seen in the fossil record, carried out initially by John Sepkoski and David Raup. They looked at the fossil record of marine animals, which are those most frequently preserved as fossils, and analysed diversity at the level of genera and families to get the broad picture. This work resulted in the now-famous graph of the number of marine families over geological time, which was first published by Sepkoski in 1981. The graph shows an increase in diversity over time from the Cambrian to the present but with five abrupt drops in diversity along the way. These occurred at or near the end of five geological periods, the Ordovician, Devonian, Permian, Triassic and Cretaceous.

Raup and Sepkoski published a paper in 1982 in which they looked explicitly at extinction rates in marine animals, and confirmed that these five declines were indeed mass extinctions. These became known as the 'big five' and later work only served to underscore their reality and importance as the five largest extinctions. The reliability of all this analysis was strengthened by the sheer size of the samples involved, with the databases running to thousands of families and tens of thousands of genera. From these data, it was possible to estimate the number of species that went extinct in each of the big five events. The numbers obtained ranged from 70% of all species at the end of the Cretaceous to 95% of all species at the end of the Permian.

These figures show the truly massive scale of these extinctions, which clearly represent global crises in the history of life. The extinction at the end of the Permian even came close to wiping out all species of animals and plants. From these studies of the fossil record, it is evident that events of this magnitude can bring about the collapse of entire ecosystems and remove previously dominant groups, such as the dinosaurs. As to the cause of these extinctions, geological evidence has indicated that the big five were not all due to one common cause. It would seem that they were caused by environmental catastrophes in various combinations, and the end-Cretaceous event is so far the only one for which there is clear-cut evidence of an asteroid impact as a key factor.

This look at extinction on the grand scale is an appropriate place at which to close our survey of research on evolution in the second half of the twentieth century. We have sampled an exciting range of studies, which have confirmed the theory of evolution in general but have also transformed the earlier understanding of many aspects. Recognition of the role played by mass extinctions in the course of evolution is one example of such a transformation. The advent of molecular biology, the continuing study of vertebrate fossils and the study of natural selection in the field have greatly enlarged our understanding of both the pattern and process of evolution. The scope and variety of all this research goes well beyond the modern synthesis, from which it started. All this may seem a trifle daunting, but it augurs well for the future of evolutionary theory.

CHAPTER TEN

EVOLUTION: TRUTH, THEORY OR MYTH?

From the outset, we have referred to evolution as a theory, and this is quite correct. Evolution is a theory. But it is as well to be clear about what this means. In science, a theory is a well thought out idea about how some part of the natural world works. The key difference between science and other fields of knowledge is that ideas in science are subject to testing by observation and experiment. A successful scientific theory is one that survives many such tests and so comes to be accepted as a reliable guide to some part of nature. Far-reaching theories, which embrace a wide range of phenomena, are obviously more important than those that are strictly limited in scope.

It is sometimes said by those unfamiliar with science that evolution is 'only a theory', implying that it is just an idle notion or at least that it is not well established. Such an idea may arise from the unfavourable comparison of theory with practice in everyday usage, but this is to misunderstand how science works. It is rather like saying to a politician that taxation is 'only a policy' or to a theologian that the atonement is 'only a doctrine'. For theories are what science is about, just as politics is about policies and theology is about doctrines. What matters is whether a theory has passed a sufficient number of tests for us to regard it as a reliable guide to how nature works. Scientific understanding is built on the progressive testing and refinement of general theories.

But do our carefully tested theories really say something true about the natural world? This point has been much debated over the years, and opinion has varied widely, as can be well imagined. At one extreme is the view that the basic theories or laws of science represent direct truths about the actual state of affairs in the natural world. On this view, science is a method of obtaining a direct read-out of the patterns and causes that are found in nature. The scientist is like an impartial spectator, who describes a game of sport and deduces the rules of play without being involved in any way himself. This way of thinking was widely adopted in the eighteenth and nineteenth centuries, when scientific confidence was at its peak, but it is rarely heard today.

The picture of the scientist as an objective spectator has died a natural death, thanks to the work of historians and philosophers of science. It is now clear that even simple observations are not imbibed passively from the external world but are made by a human mind already laden with ideas. The shaping of these ideas is a human activity carried out in a particular social context, with all the frailties and limitations that that implies. This has led some people to the other extreme, in which scientific knowledge is viewed as no more than the expression of a particular social group. On this view, there are no such things as discoveries in science, only changes in fashion about how we choose to view the world. However, such a view cannot account for the fact that scientific understanding does not merely change but is progressive. We see this most clearly in the success of technology, which must reflect an understanding of the real world and not just a change of fashion.

A sensible view of scientific theory must lie somewhere between these two extremes and embody elements of both. Certainly, scientific discovery does not involve a one-way flow of information from nature to a passive, open mind. It involves a creative interaction of mind and nature, in which scientists seek to construct an adequate picture from what they see of the world. At first, a wide range of interpretations may be possible, but in time the accumulation of new evidence from nature increasingly constrains the range of possibilities. Eventually, a time may come when one particular interpretation fits the evidence so well that it may be regarded as placed beyond reasonable doubt.

The situation is rather like one of those tests of visual perception, in which an image is reduced to a pattern of dots, and a random selection of these dots is shown to a viewer. As progressively more dots are added, the viewer tries first one and then another interpretation in an attempt to visualize the original image. In due course, the accumulation of dots constrains the possibilities to the point where the viewer feels confident of the correct interpretation.

Similarly, scientists' confidence in the theory of evolution depends on the way that it has held up as new evidence has accumulated. We have seen that evolution was convincing to Darwin's contemporaries because it enabled them to make sense of several apparently unrelated problems.

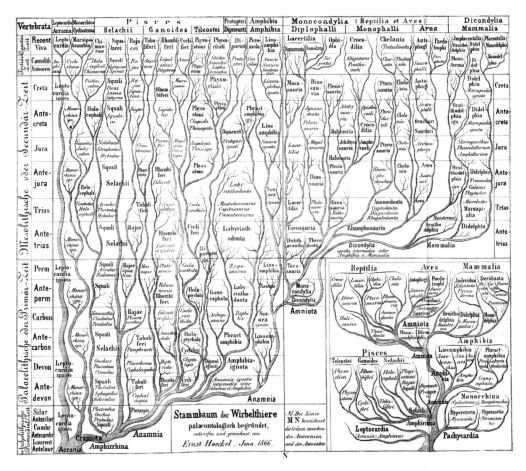

Haeckel's evolutionary tree of the vertebrates, drawn in 1866, presented a highly speculative history of the group, unsupported by fossil evidence. He got round the lack of fossil intermediates by inventing extra periods of time (Ante-carbon, Ante-devon, etc.) before each of the recognized geological periods (Carbon, Devon, etc.).

These had to do with the classification and geographical distribution of living things, the pattern of the fossil record, and the structure and embryological development of animals. With the passage of time, these areas of study continued to yield a consistent pattern of evidence that strengthened the conclusion that evolution has occurred.

Of course, if one has decided in advance either that evolution is a self-evident truth or that it must be false, then the evidence is not particularly relevant. The effect of deep prior convictions shows up well in two notable figures from the nineteenth century debates on evolution. One is Ernst Haeckel, who was enchanted by Darwin's work and quickly incorporated it in a grand philosophy of everything. He was so convinced he could deduce the course of evolution, mainly from embryology, that he virtually ignored the fossil evidence. He got around the lack of transitional forms simply by inventing extra, unrecorded periods of geological time to accommodate the hypothetical forms that his ideas required. So the

237

evolutionary trees that he drew looked convincing but they bore little relation to the fossil evidence.

By contrast, Louis Agassiz could never bring himself to accept the
theory of evolution, even though his son did. To the day he died,
Agassiz maintained that the history of life followed a preconceived plan
of creation, in which the species of each period were specially created to
suit their precise habitats. And he held out against evolution on the
grounds that there were no signs of transition between separate species
in either living or fossil forms. He did, however, make a remarkable sea
voyage late in life, on which he retraced a major part of Darwin's
Beagle voyage, as a deliberate test of the evidence for evolution. He
sailed around the coast of South America and even paid a visit to the
Galápagos, but what he saw did not alter his views.

In our journey's guide (chapter 1), it was mentioned that the theory of
evolution may be thought of as having two components. One component
deals with what has happened in the history of life on this planet. It seeks to
explain the pattern of diversity found in animals and plants by providing
an accurate account of past changes that have given rise to this diversity.
The other component deals with the mechanisms of evolution, that is the
processes of change that produce well-adapted organisms and new species.
Along our journey we have seen how, since their formulation by Darwin
and Wallace, both components of the theory have been supported by new
findings to a remarkable extent.

By way of illustration for the first component, consider the fossil
record, which provides the most direct evidence about life in the past.
When Darwin wrote the *Origin*, he could show that the overall pattern
of the fossil record supported evolution, but he was embarrassed by the
lack of intermediate forms. Since then, fossil collecting has resulted in
many fossils being found with features that are intermediate between different groups, just as the theory requires. It may be open to a deeply sceptical person to object that the gap between one fossil form and the next is
still too great to convince him of an evolutionary transition. But it is
beyond dispute that these so-called gaps in the fossil record are getting
smaller as more fossils are collected. This should not happen if evolution
were a false theory.

For example, we have seen how new fossil discoveries in the twentieth
century made it possible to document the evolutionary transition from
reptiles to mammals. The fossil record of the mammal-like reptiles in

the Permian and Triassic periods is exceptionally good and this enabled researchers from Owen and Broom onwards to trace the gradual acquisition of mammalian features. These features emerged as part of an adaptive radiation of the mammal-like reptiles, which dominated the land for many millions of years before they were displaced by the dinosaurs. The origin of the mammals as a major new group of vertebrates could thus be traced within the unfolding history of their ancestral group.

A similar story can be told from the discovery of hominid fossils. It is scarcely two centuries since Cuvier confidently asserted that 'there are no human bones in a fossil state'. Yet by the end of the twentieth century such a variety of fossil hominids were known that they needed to be classified in at least a dozen separate species. Moreover, a consistent pattern can be seen among these varied forms, even though the experts argue fiercely about the finer details. The older species were generally more ape-like in appearance but had a bipedal gait. Among the younger ones, some show progressively more human features, such as increasing cranial capacity and larger body size. Anyone who chooses to deny that these forms document an evolutionary transition from ape-like ancestors to modern humans must explain away these fossils. Since Cuvier's day, that task has become harder and harder with every passing year.

The second component of evolutionary theory, which is concerned with the mechanisms of change, was also supported by new discoveries during the twentieth century. The power of natural selection to change gene frequencies was shown by the mathematical analysis of Mendelian inheritance. This conclusion was confirmed by the rise of molecular biology, which showed just how genetic variation is generated for natural selection to work upon. In addition, a range of field studies showed how intense selection pressures could produce rapid evolutionary change in natural populations. Finally, forms of social behaviour that seemed incompatible with natural selection were analysed and shown to be a simple consequence of basic Darwinian theory.

However, the success of evolutionary theory does not mean that we have reached some haven of absolute truth where all doubts and difficulties have been removed. Aspects to be investigated now run all the way from the molecular basis of variation to major sequences of change over large periods of geological time. Given the incompleteness of our information, such a range of topics gives plenty of scope for differences of interest and of interpretation. Sometimes these different points of view are expressed in exaggerated language, particularly where the advocates of a new idea feel they are putting right an error in the prevailing view.

A good example of this is the lively debate that took place in the 1970s and 1980s over the idea of 'punctuated equilibrium'. This hypothesis was put forward by Niles Eldredge and Stephen Gould, who proposed that the rate of evolutionary change is rapid during the formation of a new species and virtually zero at other times. Hence a species remains almost unchanged in structure from the time of its first appearance until it evolves into another species or becomes extinct. They contrasted this view with a parody of the prevailing view, according to which evolution goes on steadily at a slow, constant rate of change. A close look at the fossil record, they maintained, supports the idea of punctuated equilibrium rather than the alternative, which they called 'gradualism'.

These ideas were vigorously advocated. On occasions, punctuated equilibrium was promoted not just as a healthy adjustment of the modern synthesis but as quite a new theory of evolution. It was even said that the modern synthesis 'is effectively dead'. Such claims are natural enough in some ways: they reflect an understandable tendency to emphasize the points of disagreement with one's colleagues, and they are a great way to get attention. But they can mislead those on the fringes of science into thinking that evolutionary theory is in a state of crisis, or even in terminal decline. This is not the case. There is nothing like a good fight to promote the health of a science: progress comes out of the clash of different views plus a supply of new information.

By the 1990s, new information that could put these ideas to the test had been extracted from the fossil record, through looking at well-preserved lineages. This proved to be no easy matter, but the outcome was that some lineages appear to fit the punctuated equilibrium model while others do not. Enough cases were examined to show that species often do undergo evolutionary change after their first appearance. A clear appreciation of just how variable the tempo of evolution is, and higher standards for assessing the rate of change in fossil lineages, have come out of this controversy. It has also opened up other new topics for discussion. For this is how science works: controversies such as this are a sign of health, not of decay.

One matter of general interest is the broad picture of evolution that has emerged from more than a century of study since the *Origin* first appeared. In spite of the controversies on points of detail, we have today a well-established picture of the evolution of life. This is obviously quite different from the general picture of life that prevailed when John Ray studied plants and animals in the seventeenth century. So great a change in perspective marks a significant transition in scientific thought, and it is worth reflecting briefly on the nature of this change.

During the period discussed in this book, three main pictures of life on earth have succeeded one another. The first of these might be termed 'clockwork design' because it attributed the adaptations of living things to divine design, as we saw in chapter 2. God was pictured as a sort of cosmic watchmaker, and well-adapted species were the individual clocks and watches that he had skilfully crafted when the world was created. Ray drew attention to the eye, which later haunted Darwin, and to the woodpecker as examples of adaptation that could be interpreted in this way. From the end of the eighteenth century, clockwork design began to give way as it was realized that life had a vast history, with some species becoming extinct and others replacing them.

A second picture, which might be termed 'programmed evolution', was initiated by Lamarck. In his theory, animals and plants evolve in a necessary and predictable manner, with only minor deviations. So when a new organism evolves, it already has a pre-ordained place in the scheme of things. Although Lamarck had no particular sympathy with religion, the idea of programmed evolution was interpreted in religious terms later on. Chambers, the author of the famous *Vestiges*, which appeared before the *Origin*, saw evolution as the unfolding of a divine plan. Mivart, the zoologist, took much the same view shortly after the *Origin* was published.

It was the *Origin* that introduced the third picture, which might be termed 'contingent evolution' to contrast it with the previous one. Here, variation among individuals, arising from causes internal to the organism, is recognized as a normal part of life. The nature of the variation that occurs is quite independent of the forces at work in the external environment. One of these external forces is natural selection, which is a sorting process that is limited by the available variation. Just as the breeder depends on the uncertainties of variation, Darwin explained in a letter to Lyell, 'so under nature any slight modification which <u>chances</u> to arise, and is useful to any creature, is selected or preserved'.

And the process of natural selection is itself an uncertain business, on account of the many different factors that influence the survival of an individual. Possessing a favourable variation increases the probability that an individual will survive and reproduce, but does not make it certain. As Darwin wrote in the *Origin*, an organism that carries a favourable variation 'will have the best chance of being preserved'. So evolutionary change is contingent upon the availability of suitable variation and the statistical process of natural selection, and is not the unfolding of some pre-determined plan.

In introducing these elements of uncertainty or contingency into biology, the views of Darwin and Wallace differed from those that had gone before. That evolutionary change depends on circumstances in this way, and is not pre-determined, was a new and unwelcome idea. When Sir John Herschel first read about natural selection in the *Origin*, he dismissed it as 'the law of higgledy-piggledy'. He was quite willing to accept that new species might arise by natural causes, as his correspondence with Lyell had shown. The problem was that natural selection was not the sort of natural law he was expecting. He was versed in Newtonian physics and astronomy, where there were no such uncertainties. How could such a chancy process possibly produce the well-designed organisms we see around us?

Darwin responded to this difficulty in 1862 with a short book on the *Fertilization of Orchids*. This represented 'a flank movement on the enemy', as he confided to his friend Asa Gray. In it, he showed how orchid flowers are adapted to ensure that one flower is fertilized by pollen carried from another. The way in which insects are induced to effect this cross-pollination differs from one species to another, and yet all orchid flowers are built on the same basic plan. So the same set of parts are modified to form different mechanisms of pollination in different species. On comparing the different species, it is often possible to find intermediate stages, by which one mechanism could have evolved into another. 'The regular course of events seems to be,' Darwin concluded, 'that a part which originally served for one purpose, by slow changes becomes adapted for widely different purposes.'

Hence the way in which each species has solved the problem of cross-pollination seems to have depended on the circumstances. When a variation has arisen that leads to a species solving the problem in one way, the modified parts create the opportunity for another solution to evolve, and so on. The adaptations found in a given species are not so much well designed as cobbled together from the parts available in its immediate

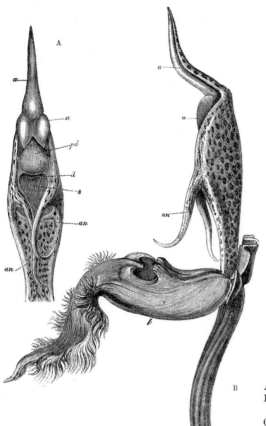

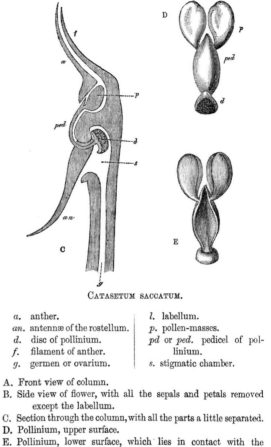

CATASETUM SACCATUM.

a. anther.	*l.* labellum.
an. antennæ of the rostellum.	*p.* pollen-masses.
d. disc of pollinium.	*pd* or *ped.* pedicel of pol-
f. filament of anther.	linium.
g. germen or ovarium.	*s.* stigmatic chamber.

A. Front view of column.
B. Side view of flower, with all the sepals and petals removed except the labellum.
C. Section through the column, with all the parts a little separated.
D. Pollinium, upper surface.
E. Pollinium, lower surface, which lies in contact with the rostellum.

CATASETUM SACCATUM.

Male flower of the orchid, Catasetum, *from Darwin's book on the* Fertilization of Orchids. *LEFT, drawing of the flower with all the sepals and petals removed except the labellum (*l*). RIGHT, diagram of the same region to show the relationship of the various parts. Darwin was able to show how the pollen masses (*p*) become attached to a visiting insect by means of the adhesive disc (*d*).*

ancestors. As Darwin remarked, it is rather as 'if a man were to make a machine for some special purpose, but were to use old wheels, springs, and pulleys, only slightly altered'. The same kind of evidence for contingency can now be obtained from almost any group of animals or plants that has been well studied. Tree kangaroos, which were discussed in chapter 1, are a good case in point.

An additional level of contingency, of which Darwin was not aware, is introduced by events on the grand scale, such as mass extinctions. When the dinosaurs died out at the end of the Cretaceous, for instance, the final blow was almost certainly delivered by a large asteroid colliding with the Earth. So the dinosaurs were not driven to extinction because they were less well adapted than other land animals, they were wiped out by a global catastrophe. In the words of the palaeontologist David Raup, it was a case of their having bad luck rather than bad genes.

In turn, the demise of the dinosaurs created an opportunity for some remarkable animals to evolve in their place. The famous carnivorous

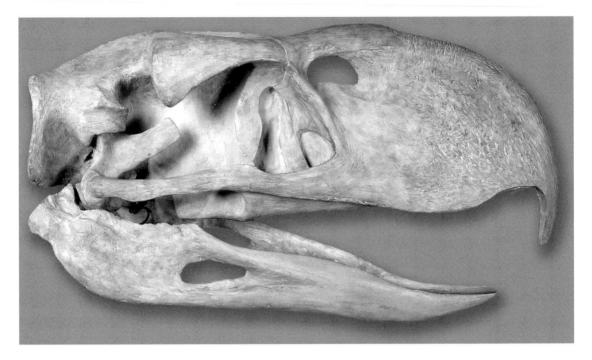

dinosaurs were replaced in due course by new carnivores that included not only mammals but also large, flightless birds. In the Eocene, the genus *Diatryma* appeared and ranged over Europe and North America for several million years. This bird stood over 2 m tall, with powerful legs but vestigial wings, and had a large head with a huge, curved beak. It was evidently a fierce carnivore.

It is fascinating that a similar but unrelated group of birds evolved independently in South America, which was then still an island. These, too, were large ground birds, belonging to the genus *Phorusrhacus* and its relatives, and they were more successful than their North American counterparts. They flourished alongside the distinctive mammalian carnivores of South America for a long time, and only became extinct a few million years ago. However, these two groups of giant predatory birds would probably never have existed were it not for the demise of the dinosaurs.

We can conclude that the picture of contingent evolution, of opportunistic progress through time, has been upheld by the results of science since Darwin. The contingent aspect of evolution is not just some speculative concept but is impressed on us by the plain facts of natural history. Such a pattern does make sense if genetic variation and natural selection are important in evolution. For then the direction of evolution will depend on the available variation at a given time and place, and the

Skull of the giant ground-dwelling bird, Phorusrhacus, *one of a family of birds that flourished in South America after the demise of the dinosaurs. The skull is some 60 cm in length.*

selection pressures operating at that particular time and place. From all this, it would appear that evolution is a unique historical process. As far as we can see, it really could have taken a different course from the one that has actually occurred.

So where does that leave us? Maybe our own species would not have evolved if all those dinosaurs had not been dealt a mortal blow. Are we then just the product of an historical accident, a throw of the cosmic dice? Few people can remain indifferent to such questions since they touch on our status in the universe at large and hence on the meaning or lack of it in our existence. For that very reason, it is advisable to tread carefully in attempting to deal with these questions. The key point to bear in mind is that in asking about meaning or purpose we are moving out of science, which deliberately tries to exclude such notions, and into philosophy and theology.

This transition from science to philosophy stands out clearly in the argument from design, which prevailed in the seventeenth and eighteenth centuries. From their scientific work, naturalists such as Ray recognized adaptation as an important fact of life, but they also saw it as evidence for the existence of God. Given the Christian perspective of their day, there seemed to be no problem in moving straight from the scientific observation of adaptation to a philosophical conclusion about the universe as a whole. The results of biological science were taken at their face value as a clue to the ultimate meaning of the world and our place in it. However, the argument from design lost ground in the nineteenth century, when it was found that adaptation could be explained in scientific terms, by natural causes acting over vast periods of time.

The failure of the design argument does not seem to have made people more cautious about the whole business of jumping from scientific results to philosophical conclusions. As a rule, they have been content merely to jump to different conclusions. One approach among biologists has been to take it for granted that the contingency of the evolutionary process provides a simple and direct clue to the ultimate meaning of the universe. This contingency shows that organisms are not designed, nor is the course of evolution planned, in any ordinary sense. Hence we must accept that the universe is purposeless and our existence meaningless, so the argument runs.

This shift from scientific interpretation to grand philosophy is made quite explicit in *The Meaning of Evolution*, published by George

Simpson in 1949. 'Man is the result of a purposeless and materialistic process that did not have him in mind', he wrote, and he added that 'the universe apart from man or before his coming lacks and lacked any purpose'. A number of other biologists have continued to say much the same thing down to the present day. For example, Simpson's thesis was restated near the end of the twentieth century by Stephen Gould in *Wonderful Life*, which uses the Cambrian fossils from the Burgess Shale to illustrate the contingency of evolution. An inevitable consequence of this contingency, we are told, is that our ultimate status is one of 'increasing marginality in an uncaring universe'.

But this simply will not do. Major questions of philosophy cannot be settled in so cavalier a fashion. We need first of all to ask what the limits of evolutionary biology are and whether it is legitimate to extrapolate its observations and theories way beyond their original scope. Can a rational interpretation of the universe as a whole be reached simply by generalizing without limit the concepts of one scientific discipline? Without such critical enquiries, our philosophical conclusions will be based on the same sort of naive extrapolation of biology as the argument from design. As it stands, therefore, the philosophies advocated by some biologists have little enough foundation, but they tend to acquire a spurious authority by being linked with sound scientific exposition.

All this may seem rather far removed from the details of natural history, which have been the main focus of this book. However, as far as evolution is concerned, the link between the details of natural history and these larger issues has been there from the beginning. As soon as the *Origin* was published, thoughtful people began to ponder its significance for such topics as the status of human beings and the theological concept of creation. Nor can these concerns be avoided, for the theory of evolution does touch on some of the profound issues that confront us as human beings, as we have noted. So it is perfectly natural that the link with philosophy and theology should hold a special fascination for us, as it did for both Darwin and Wallace.

This is obviously not the place to pursue such issues, but it may be appropriate to urge both caution and moderation. Caution is needed because the risk of jumping to unjustified conclusions gets greater as we widen the scope of discussion beyond our own particular speciality. Most of us today are limited by having been educated in only one discipline, and our area of special competence is usually narrower still.

"But then arises the doubt, can the mind of man, which has, as I fully believe, been developed from a mind as low as that possessed by the lowest animal, be trusted when it draws such grand conclusions?" (C. Darwin, 'Autobiography', 1876)

Evolution depicted in a Victorian cartoon, used as the frontispiece for a book by David Lack in 1957. He coupled this with a quotation from Darwin, expressing a cautious attitude to grand conclusions.

This makes it hard to achieve a reliable discussion of issues that range across the boundaries between biology, philosophy and theology.

Moderation is also an asset in these discussions. The complete separation of different disciplines, which has been produced by generations of professional specialization, leads not only to mutual ignorance but also to mutual hostility. The practitioners in each discipline may be inclined to draw far-reaching conclusions solely from the results of their own area and to ignore the others entirely. Tempting though such short cuts may be, they are unlikely to yield a satisfying result in the long term. A more moderate approach, in which people from different disciplines are willing to listen to each other, seems more likely to resolve some of the profound questions that face us.

That it is worth resolving these questions is becoming more apparent, in a way that was not evident a generation or two ago. Until quite recently, it was unthinkable for those who wrote on evolution in the grand style to suggest that our future could be in doubt. On the contrary, evolution was used to underwrite all sorts of marvellous prospects for the human race. But now things look different, and there is a pressing need to secure the future of our species and that of the other species with which we share this planet. To do this, a better knowledge of evolutionary biology will be needed, but so too will religious values and wise philosophy. It is precisely in the links between the separate disciplines of human knowledge that the crucial issues lie. If we cannot get together to work them out, we too may join the long list of extinct species.

But let us finish as we began, with the quest to understand the living world. It was the fascinations of natural history that led people of earlier centuries to the discovery of evolution. Today, the great strength of evolution as a theory is still its ability to throw light on the current problems of research in biology. Things such as fossil lineages, DNA sequences and patterns of animal behaviour are the testing ground of new developments in the theory. In the long run, elegant speculation or powerful rhetoric are no substitute for this grounding in facts gathered in the field or in the laboratory. The modern equivalent of natural history still stands, therefore, as the basis of the contribution that the theory of evolution can make to human understanding.

EVOLUTIONARY READING

This reading list offers a modest selection of books for anyone who wants to follow up aspects of evolution covered in our narrative. A good textbook on evolution from a history of science perspective is Bowler (2003). A valuable discussion of events in the nineteenth century is also provided by Ruse (1979/1999), and Secord (2000) paints an evocative portrait of debate just before the *Origin of Species*. Much useful information about the history of biology is also available in Mayr (1982). All of these history books have full bibliographies, which makes it possible to follow up particular points in more detail.

The importance of the debates about fossils is explained by Rudwick (1976), and Cohen (2002) explores the impact of the mammoth on these debates. Similarly, the role played by work on the geographical distribution of plants and animals is made clear by Browne (1983). These three books have the advantage that they carry the narrative back well before the nineteenth century. These earlier times are somewhat neglected but Toulmin and Goodfield (1965) is a help here, as is Thomas (1983) from a different perspective. The eighteenth-century naturalists, from Linnaeus to Buffon, are considered by Larsen (1994). British naturalists from this period are included in Porter (2000). The discovery of geological time was a prerequisite for the theory of evolution, and this story is told in a series of biographical sketches by Albritton (1980) and in a charming essay by Gould (1987). See also some important essays in the volume edited by Lewis and Knell (2001).

A number of books pay special attention to developments after Darwin's lifetime. Bowler (1996) looks at the attempts to reconstruct an evolutionary history of life, and the essays edited by Mayr and Provine (1980/1998) review the formation of the modern synthesis. Cronin (1991) follows with zeal the developments that have enabled social behaviour in animals to be explained in terms of natural selection.

Reading history books is no substitute for reading the original work of biologists of an earlier period. Possibly the best introduction to nineteenth-century writing is to read the accounts of their travels in the cause of natural history written by Bates (1863), Darwin (1845) and Wallace (1863), all of which have been reprinted as modern paperbacks. Darwin's *Origin of Species* is also available in modern reprints, both of the first edition (1859) and of the sixth (1872). For most other works, a visit to a long-established library is called for; among the most readable of the older books to be found there are those of Wallace (1889, 1891).

On coming to developments in the latter part of the twentieth century, we encounter a vast literature. Fortunately, there are some excellent textbooks on evolution, which provide a broad coverage of the subject, such as Futuyma (2005) and Ridley (2003), both of which have ample lists of further reading. A good introduction to the contribution that developmental biology is making to our understanding of evolution is given in Carroll (2005).

For an account of fossils and their interpretation in relation to evolution, a good place to start is Fortey (2002) along with Kemp (1999). The fossil history of vertebrates is covered in readable form by Colbert (2001) and Benton (2004), and selected issues are explored by Carroll (1997). The origin of mammals in particular is considered in detail by Kemp (2004), and an introduction to the study of hominid fossils is provided by Lewin (2004), Stringer and Andrews (2005) and Tattersal (1995).

A discussion of natural selection based on field evidence will be found in Endler (1986) and for details of Darwin's finches studied on the Galápagos islands, see Grant (1986/1999) and Lack (1947/1983). The successful development of natural selection theory to cover social behaviour of animals is introduced by Krebs and Davies (1987) and dealt with in more detail by Krebs and Davies (1991). On the topic of extinctions, a good starting point is Raup (1991), with more details on mass extinction in Hallam and Wignell (1997) and Hallam (2004). A good scientific account of the dinosaurs and their extinction is provided by Fastovsky and Weishampel (2005).

Finally, the late twentieth century debate over the concept of punctuated equilibrium can be sampled in several places. Enthusiastic support can be found in Eldredge (1986) and Stanley (1981), and later, more critical accounts in Gould (2002), Kemp (1999) and Levington (2001).

EVOLUTIONARY WHO'S WHO

This is a biographical list, in alphabetical order, of those people whose work is included in the preceding narrative. Authors who were still alive at the time of writing are not included.

AGASSIZ, JEAN LOUIS RODOLPHE (1807–73) Swiss naturalist famous for his geological work on glaciers and the resulting theory of the 'Ice Age'. Also made a special study of living and fossil fishes. Moved to the United States in 1847, where he subsequently became the leading opponent of Darwin's theory of evolution.

ARISTOTLE (384–22 BCE) Greek philosopher, tutor to Alexander the Great. A quarter of his published work was devoted to zoology; he mentioned some 500 species, and thought deeply about general problems of biology.

BATES, HENRY WALTER (1825–92) Naturalist who travelled with Wallace to the tropics of South America. There he collected over 8000 species of insects new to science. He was the first to explain those close external resemblances between different species which are now known as 'Batesian mimicry'.

BATESON, WILLIAM (1861–1926) Biologist at the University of Cambridge, who made a great contribution to establishing the Mendelian concept of heredity and coined the word 'genetics'. He supported the theory of evolution but denied the importance of natural selection.

BLUMENBACH, JOHANN FRIEDRICH (1752–1840) Professor of medicine at the University of Gottingen; he devoted much of his time to natural history, especially fossils and geology. These topics were discussed in his *Handbook of Natural History*, where he drew conclusions similar to those of Cuvier.

BOUCHER DE PERTHES, JACQUES (1788–1868) French customs officer who took up archaeology. His careful study of flint tools in the Somme river valley helped to establish the fact that humans had once coexisted with extinct species of mammals.

BRONGNIART, ADOLPHE-THEODORE (1801–76) Son of Alexandre' Brongniart, and an able botanist. His main work concerned the distribution of fossil plants through time and showed that there were progressive changes in the fossil species.

BRONGNIART, ALEXANDRE' (1770–1847) Eminent geologist who worked with Cuvier to determine the order of the rock formations around Paris. Like William Smith, he developed the use of fossils in correlating a series of strata over a large area.

BRONN, HEINRICH GEORG (1800–62) Professor of natural science at the University of Heidelberg, his special study was the distribution of fossil animals through time. His prize-winning essay on the subject showed that the changes in fossil species had been gradual and progressive.

BROOM, ROBERT (1866–1951) Born in Scotland and trained as a doctor, he settled in South Africa after a brief period in Australia. There he devoted himself to studying fossils of the mammal-like reptiles and to finding and describing the oldest human fossils.

BUCKLAND, WILLIAM (1784—1856) The first person appointed to teach geology at the University of Oxford, he promoted this new science with great energy and enthusiasm. Ichthyosaurs and the geology of caves were his special topics of study. Noted for his successful Bridgewater treatise on *Geology and Mineralogy*.

BUFFON, GEORGES LOUIS LECLERC, COMTE DE (1707—88) A scientist with wide interests and great influence, he was director of the King's Garden in Paris for most of his career. His work at the Garden resulted in his monumental *Natural History*, which ran to 36 volumes and explored topics such as the age of the earth and the possibility of evolution.

BURNET, THOMAS (1635—1715) Had a long association with the University of Cambridge, and was for a time chaplain to King William III. He was forced to resign the latter position through the controversy that followed publication of his famous book, *The Sacred Theory of the Earth*.

CARPENTER, WILLIAM BENJAMIN (1813—85) Medical professor at the University of London, who published wide-ranging work in both physiology and zoology. He was one of the first people to write a favourable review of Darwin's *Origin of Species*, although he never fully accepted natural selection.

CASTLE, WILLIAM ERNEST (1867—1962) A professor of biology at Harvard University, with a life-long interest in the study of evolution. After 1900, he spent most of his time on studies of the relationship between Mendelian inheritance and evolution.

CHAMBERS, ROBERT (1802—71) Founded the Chambers publishing house, with his brother, in 1832. His interest in science led him to publish the *Vestiges of the Natural History of Creation* in 1844. This book stirred up public debate on evolution in the years before Darwin's *Origin of Species*.

CONYBEARE, WILLIAM DANIEL (1787—1857) English geologist and theologian, closely associated with Buckland at Oxford. He wrote one of the best early textbooks on geology (with W. Phillips), and described the fossil *Plesiosaurus*. He was a penetrating critic of Lyell's *Principles of Geology*.

COPE, EDWARD DRINKER (1840—97) Pioneer in the study of vertebrate fossils in the United States. He collected and described many important new finds, and clashed with Marsh in this activity. He was a leading advocate of the Lamarckian interpretation of evolution in the late nineteenth century.

CUVIER, GEORGES LEOPOLD CHRETIEN FREDERIC DAGOBERT (1769—1832) Professor of animal anatomy at the Museum of Natural History in Paris. With his exceptional skill in comparative anatomy, he made crucial contributions to the classification of animals and the reconstruction of fossil vertebrates. He firmly rejected the evolutionary theory of his colleague Lamarck.

DARWIN, CHARLES ROBERT (1809—82) English naturalist famous for his book on the *Origin of Species*, in which he set out the theory of evolution by natural selection, based on 20 years of study. He began to form his ideas on evolution after five years as a naturalist on board HMS *Beagle*.

DARWIN, ERASMUS (1731—1802) Doctor, inventor and grandfather of Charles Darwin. Among his books on natural history was *Zoonomia*, in which he explicitly advocated the idea of biological evolution.

DAUBENTON, LOUIS JEAN-MARIE (1716—1800) Trained in medicine, Daubenton used his anatomical skills to assist Buffon in the producing the latter's *Natural History* series. In doing so, he was a pioneer of comparative anatomy. Eventually became the first director of the Museum of Natural History in Paris.

DESMAREST, NICOLAS (1725–1815) French scientist who concentrated increasingly on geology. He made a detailed study of the extinct volcanoes in the Auvergne district, and concluded that all basalt is volcanic in origin.

DE VRIES, HUGO (1848–1935) Dutch botanist whose experiments on heredity led to the rediscovery of Mendel's work and to the recognition of mutation. As part of his mutation theory, he proposed that new species arise suddenly through mutation rather than gradually through natural selection.

DOBZHANSKY, THEODOSIUS (1900–75) Geneticist who began his career in Russia and migrated to the United States in 1927. He specialized in research on wild populations of the fruit fly *Drosophila*. His book on *Genetics and the Origin of Species* played a crucial part in bringing about the modern synthesis of evolutionary theory.

FISHER, RONALD AYLMER (1890–1962) English statistician and geneticist. He demonstrated that the Mendelian concept of heredity is compatible with blending inheritance and natural selection. This work was brought together in his *Genetical Theory of Natural Selection*, which paved the way for the modern synthesis of evolutionary theory.

GALTON, FRANCIS (1822–1911) A student of biological variation and cousin of Charles Darwin. He pioneered the application of statistical methods to heredity and variation, and was keenly interested in the relation of heredity to evolution.

GAUDRY, ALBERT JEAN (1827–1908) Associated with the Museum of Natural History in Paris from an early age, he made a special study of fossil mammals and reptiles. Following publication of Darwin's *Origin of Species*, he was one of the first to use specific fossils to construct evolutionary trees.

GEOFFROY SAINT-HILAIRE, ETIENNE (1772–1844) A professor of zoology at the Museum of Natural History in Paris. His skill in anatomy and embryology enabled him to demonstrate a 'unity of plan' (homology) in the structure of vertebrates. He disagreed with Cuvier's conclusions and sympathized with the evolutionary ideas of Lamarck.

GOODRICH, EDWIN STEPHEN (1868–1946) The leading comparative anatomist of his day, becoming professor of zoology at the University of Oxford. He used embryological studies to clarify the problems of homology, and was always interested in evolutionary studies.

GRAY, ASA (1810–88) Gave up a career in medicine to study botany, and became the leading classifier of plants in the United States. He met Darwin at Kew in 1851 and they kept up a regular correspondence. Later on he accepted Darwin's theory of evolution and sought to reconcile it with religion.

GULICK, JOHN THOMAS (1832–1923) Son of an American missionary in Hawaii, he made a huge collection of Hawaian snails at the age of twenty-one. He spent the rest of his life analysing this collection in the breaks from his own missionary work, and concluded that geographical isolation is important in the evolution of new species.

HAECKEL, ERNST HEINRICH PHILIPP AUGUST (1834–1919) German zoologist with a special interest in marine invertebrates. After 1860, he became one of Darwin's most enthusiastic advocates, and introduced many useful concepts into biology. He also promoted evolution as a complete philosophy that could replace traditional religion.

HALDANE, JOHN BURDON SANDERSON (1892–1964) A remarkable biologist who did research in physiology, biochemistry and genetics, as well as writing popular science. He wrote a series of mathematical papers on the Mendelian basis of natural selection and a book on *The Causes of Evolution*.

HENSLOW, JOHN STEVENS (1796–1861) Professor of botany at the University of Cambridge, where he gave a popular series of lectures. He helped to promote science in many ways. As an undergraduate at Cambridge, Darwin became known as 'the man who walks with Henslow'.

HERSCHEL, JOHN FREDERICK WILLIAM (1792–1871) A famous astronomer and son of a famous astronomer, he was a major public figure in British science. His *Preliminary Discourse* discussed the nature of science in general and strongly influenced Darwin.

HOOKE, ROBERT (1635–1703) English inventor and scientist, founder member of the Royal Society. His most important book was the *Micrographia*, in which he not only revealed the minute features of insects but also began the study of fossils and discussed geology.

HOOKER, JOSEPH DALTON (1817–1911) Like his father, Sir William, a distinguished botanist and Director of the Royal Botanic Gardens at Kew. He specialized in classification and plant geography, and the evidence from these areas made him a convinced supporter of Darwin's views from 1860 onwards.

HUTTON, JAMES (1726–97) Scottish naturalist noted for his vital contribution to early geology. His *Theory of the Earth* was the first work to envisage cycles of erosion and uplift in the formation of the land and to recognize the igneous origin of granite.

HUXLEY, THOMAS HENRY (1825–95) Now remembered mainly for his vigorous defence of Darwin, he was a zoologist who published work on vertebrates, invertebrates and fossils. His later works were devoted to evolution and to scientific education.

JOHANNSEN, WILHELM LUDVIG (1857–1927) Danish biologist who strongly influenced genetics in its formative years. He coined the word 'gene' and drew the distinction between 'genotype' and 'phenotype'. His main research interest was the inheritance of characters that vary quantitatively.

JORDAN, DAVID STARR (1851–1931) An energetic American biologist who specialized in the study of fishes, describing over 2000 species. He strongly supported and contributed to a Darwinian view of evolution.

KELLOGG, VERNON LYMAN (1867–1937) A biologist who specialized in the study of insects, becoming professor of entomology at Stanford University. He also wrote several books on evolution, one with D. S. Jordan.

KIRCHER, ATHANASIUS (1602–80) Born and educated in Germany, he spent most of his career in Rome. He wrote many books on a vast range of subjects including science; his interest in Noah's Ark formed only a small part of his total output.

LACK, DAVID LAMBERT (1910–73) Began as a school teacher in Devon and went on to become director of the Edward Grey Institute of Ornithology in Oxford. His outstanding field work on birds enabled him to make important studies in the ecology and evolution of animal species.

LAMARCK, JEAN BAPTISTE, CHEVALIER DE (1744–1829) French naturalist with wide theoretical and philosophical interests. He distinguished himself in botany and later in zoology at the Museum of Natural History in Paris. It was there that he developed a theory of evolution, set out fully in his *Zoological Philosophy*.

LINNAEUS, CARL (1707–78) Oustanding botanist, renowned for establishing a new system for classifying plants and animals, in his *System of Nature*. In later editions, he introduced the method of giving each species two Latin names. This work formed the foundation of classification in biology.

LYELL, SIR CHARLES (1797–1875) Author of the brilliant *Principles of Geology*, which he updated through many editions. Turned from the law to a career in geology after his interest was aroused by Buckland's lectures at Oxford. Later became a close friend of Darwin and eventually supported the theory of evolution.

MALTHUS, THOMAS ROBERT (1766–1834) Began his career at Cambridge and later left to become a professor of political economy. He wrote a major textbook on the subject, but it was his *Essay on the Principle of Population* that influenced both Darwin and Wallace.

MARSH, OTHNIEL CHARLES (1831–99) A rival to Cope in exploring the many fossil discoveries of the western United States, he became professor of palaeonotology at Yale University. Best known for his work on the evolution of the horse, but he also studied the evolution of birds and dinosaurs.

MAYR, ERNST (1904–2005) Began his career in Germany and moved to the United States in 1930, eventually becoming professor of zoology at Harvard University. His special study was species formation, based on extensive field work. He was a pivotal figure in forging the modern synthesis of evolutionary theory with his book on *Systematics and the Origin of Species* and later *Animal Species and Evolution*.

MENDEL, JOHANN GREGOR (1822–84) Laid the basis of genetics with his experimental work in plant breeding. He published just two papers on the subject, and these were overlooked until 1900, when his work was rediscovered and its significance appreciated.

MIVART, ST. GEORGE JACKSON (1827–1900) English zoologist with a sound reputation in vertebrate anatomy. In his *Genesis of Species*, he accepted evolution but gave all the main reasons for doubting the importance of natural selection.

MORGAN, THOMAS HUNT (1866–1945) A major figure in early genetics, who received a Nobel Prize for his work in establishing the fact that genes are located in a line along the chromosomes. Also wrote books on the relationship between heredity and evolution.

MURCHISON, RODERICK IMPEY (1792–1871) A landowner's son who took up geology through the influence of Buckland. His research on the succession of rocks in England and Wales, and later Russia, helped to establish the periods now recognized as divisions of the geological time scale.

OSBORNE, HENRY FAIRFIELD (1857–1935) Studied vertebrate fossils at the American Museum of Natural History for most of his career. In addition to his detailed descriptions of vertebrate fossils, he wrote about the wider problems of evolution and coined the term 'adaptive radiation'.

OWEN, SIR RICHARD (1804–92) A leading comparative anatomist, who became superintendent of the natural history departments at the British Museum. His studies on the skeletons of living and fossil vertebrates were of major importance for later work on evolution. He accepted evolution in general but not natural selection.

PICTET, FRANCOIS JULES (1809–72) Fossil specialist and author of the respected four-volume *Treatise on Palaeontology*, which influenced Wallace. His review of the *Origin of Species* was, in Darwin's opinion, the only one that was opposed and yet perfectly fair. Subsequently, Pictet accepted the theory of evolution.

PLOT, ROBERT (1640–96) A clergyman whose scientific interests led to his being elected a Fellow of the Royal Society. He published a book on the *Natural History of Oxfordshire*, and became the first curator of Oxford's Ashmolian Museum, as well as being a professor at the University.

POULTON, EDWARD BAGNELL (1856–1943) Zoologist who became a professor at the University of Oxford. His special study was protective coloration in insects as an example of evolutionary adaptation. He supported the theory of natural selection and opposed the mutation theory of de Vries.

RAY, JOHN (1627—1705) English naturalist who pioneered the scientific classification of plants and, with his friend Willughby, animals. He thought out how to define a species, and produced many useful catalogues of species, notably his *Synopsis* of the British flora. In other books, he discussed topics such as the argument from design and the significance of fossils.

SCROPE, GEORGE JULIUS POULETT (1797—1876) Took up geology, with a special interest in volcanoes, after seeing an eruption of Mount Vesuvius. His study of the extinct volcanoes in Auvergne gave evidence of geological processes acting gradually over vast periods of time and had a significant influence on Lyell.

SEDGWICK, ADAM (1785—1873) Professor of geology at the University of Cambridge for over 50 years and an outstanding field geologist. His work was important in establishing the named sequence of geological periods. He was a friend of Darwin, but never accepted evolution.

SIMPSON, GEORGE GAYLORD (1902—84) Fossil specialist who became Agassiz professor of vertebrate palaeontology at Harvard University. His main research was to work out the evolutionary history of early mammals, while his books on evolution helped to forge the modern synthesis in the 1940s and 1950s.

SMITH, WILLIAM (1769—1839) Consultant surveyor who discovered that particular rock strata often contain a characteristic assemblage of fossils that can be used to distinguish them from other strata. He used this discovery to prepare the first geological map of the whole of England, and so founded the geological technique known as stratigraphy.

STENO, NICOLAUS; originally NIELS STENSEN (1638—86) Scientist noted for his work in anatomy and also in geology. One of the first people to recognize the true nature of fossils and to appreciate their significance as evidence about the past. He wrote the *Prodromus*, a brilliant discussion of fossils and sedimentation, which provided a foundation for later geology.

SUMNER, FRANCIS BERTODY (1874—1945) American naturalist whose deep interest in evolution led him to make a long-term study of wild populations of the deermouse *Peromyscus*. This work contributed significantly to the understanding of subspecies and species formation.

TOPSELL, EDWARD (1572—1638) A clergyman with many interests, chief among them being natural history. His finely illustrated *Historie of Four-footed Beasts* shows the beginnings of a scientific approach to natural history, but also shows much deference to the authority of classical sources.

TYSON, EDWARD (1650—1708) Trained at the University of Oxford, he established a medical practice in London. There he built a reputation as a comparative anatomist and was elected a Fellow of the Royal Society. Among the fine anatomical descriptions he published through the Society were those of a porpoise, a rattlesnake and a chimpanzee.

VON BAER, KARL ERNST (1792—1876) Born in Estonia, he became famous for his work in embryology carried out at the University of Konigsberg in Germany. He discovered the mammalian egg, and published a thoughtful analysis of the whole of vertebrate development from conception to birth.

WAGNER, MORITZ (1813—87) German naturalist and explorer, who travelled widely in Africa, Asia, and in North and South America. He paid special attention to the geographical distribution of animals and plants, with a view to working out how new species are formed in evolution.

WALLACE, ALFRED RUSSEL (1823—1913) Thought out the idea of natural selection as a mechanism of evolution independently of Darwin, while exploring in Indonesia and Malaysia. After the joint publication with Darwin in 1858, he continued to do original work on evolution. His studies on animal coloration and geographical distribution were specially important.

WEISMANN, AUGUST FRIEDRICH LEOPOLD (1834—1914) Professor of zoology at the University of Freiburg for most of his career. He used his knowledge of cell research to argue that the inheritance of acquired characters is virtually impossible. Hence he was an outspoken critic of Lamarck and a firm supporter of Darwin and Wallace.

WERNER, ABRAHAM GOTTLOB (1749—1817) Based at the mining academy at Freiburg for over 40 years, he was the leading geologist of his day. He recognized that stratified rocks occur in a definite order of succession and developed a rock classification based on this historical succession.

WHEWELL, WILLIAM (1794—1866) A notable figure in Victorian science, his entire career was spent at Trinity College, Cambridge. His interests were broad and even extended to coining new scientific terms, including the word 'scientist'. Later in life, he worked mainly in the history and philosophy of science.

WOODWARD, JOHN (1665—1728) Although trained somewhat haphazardly in medicine, his main interest became geology after he began collecting fossils as a young man. He defended the view that fossils really are the remains of once-living animals and plants, and supposed that they had been entombed during a universal flood.

WRIGHT, SEWELL GREEN (1889—1988) Began his career as a student of William Castle, and went on to develop the mathematical study of evolutionary change in terms of Mendelian genetics. Also collaborated with Dobzhansky on the genetics of natural populations. His work formed an important part of the modern synthesis of evolutionary theory.

BIBLIOGRAPHY

AGASSIZ, L. (1833—43) *Recherches sur les Poissons Fossiles.* 5 vols. Neuchatel.

AGASSIZ, L. (1840) *Etudes sur les Glaciers.* Neuchatel. (English translation and introduction by A. V. Carozzi; Hafner, New York, 1967.)

AGASSIZ, L. (1842) On the succession and development of organised beings at the surface of the terrestrial globe, being a discourse delivered at the inauguration of the Academy of Neuchatel. *Edinburgh New Philosophical Journal* 23: 388—99.

AGASSIZ, L. (1844—5) *Monographie des Poissons Fossiles du Vieux Gres Rouge ou Systeme Devonian (Old Red Sandstone) des Iles Britanniques et de Russie.* Neuchatel.

ALBRITTON, C. C. JR (1980) *The Abyss of Time. Changing Conceptions of the Earth's Antiquity after the Sixteenth Century.* Freeman Cooper, San Francisco.

ALVAREZ, L. W., ALVAREZ, W., ASARO, F., & MICHEL, N. V., (1980) Extraterrestrial cause for the Cretaceous—Tertiary extinction: experimental results and theoretical interpretations. *Science* 208: 1095—1108.

BAER, K. E. VON (1828) *Uber Entwicklungsgeschichte der Thiere, Beobachtung und Reflexion.* Borntrager, Konigsberg.

BAKEWELL, R. (1813) *An Introduction to Geology.* London.

BARBER, R. (1993) *Bestiary: an English version of M S Bodley 764.* (Translated and introduced by R. Barber; Boydell Press, London).

BATES, H. W. (1862) Contributions to an insect fauna of the Amazon Valley. *Transactions of the Linnean Society of London* 23: 495—566.

BATES, H. W. (1863) *The Naturalist on the River Amazons.* 2 vols, John Murray, London (4th edition, 1876; reprinted in paperback, Dover Publications, 1975).

BATESON, W. (1894) *Materials for the Study of Variation Treated with Especial Reference to Discontinuity in the Origin of Species.* Macmillan, London.

BATESON, W. (1909) *Mendel's Principles of Heredity.* Cambridge University Press. (Reprinted with additions, 1913.)

BELON, P. (1555) *L'Histoire de la Nature des Oyseaux.* Paris.

BENTON, M. J. (2004) *Vertebrate Palaeontology.* 3rd edition. Blackwell Science, Oxford.

BEWICK, T. (1885) *Thomas Bewick's Works, Vol. III, History of Quadrupeds.* Memorial edition. Bernard Quaritch, London.

BLUMENBACH, J. F. (1790) *Beytrage zur Naturgeschichte.* Gottingen. (2nd edition, 1806).

BOWLER, P. J. (1996) *Life's Splendid Drama: Evolutionary Biology and the Reconstruction of Life's Ancestry, 1860—1940.* University of Chicago Press.

BOWLER, P. J. (2003) *Evolution: the history of an idea.* 3rd edition, University of California Press.

BROOM, R. (1932) *The Mammal-like Reptiles of South Africa.* Witherby, London.

BRONGNIART, A. (1829) General considerations on the nature of the vegetation which covered the earth at the different epochs of the formation of its crust. *Edinburgh New Philosophical Journal* 6: 349—71.

BRONN, H. G. (1861) Essai d'une reponse a la question de prix proposee en 1850 par l'Academie des

Sciences etc. *Supplement aux Comptes Rendus des Seances de l'Academie des Sciences* 2: 377–918. (First published in 1858, in German, at Stuttgart).

BROWNE, J. (1983) *The Secular Ark. Studies in the History of Biogeography.* Yale University Press, New Haven.

BUCKLAND, W. (1823) *Reliquiae Diluvianae. Or Observations on the Organic Remains Contained in Caves, Fissures, and Diluvial Gravel, and on other Geological Phenomena, Attesting the Action of an Universal Deluge.* John Murray, London.

BUCKLAND, W. (1824) Notice on the Megalosaurus, or Great Fossil Lizard of Stonesfield. *Transactions of the Geological Society of London* 1: 119–30.

BUCKLAND, W. (1836) *Geology and Mineralogy Considered with Reference to Natural Theology.* 2 vols.; Bridgewater treatise No. 6. William Pickering, London.

BUFFON, G. L. L., COMTE DE (1749–67) *Histoire naturelle, generale et particuliere.* 15 vols. Paris.

BUFFON, G. L. L., COMTE DE (1778) *Les Epoques de la Nature.* Supplement to the fifth volume of the *Histoire naturelle*, Paris. (Modern reprint edited by J. Roger, Museum national d'Histoire naturelle, Paris, 1962).

BUFFON, G. L. L., COMTE DE (1792) *Barr's Buffon: Buffon's Natural History containing a theory of the Earth.* 10 vols. J. S. Barr, London.

BURNET, T. (1690) *The Sacred Theory of the Earth.* 2 vols. Walter Kettilby, London. (3rd edition, 2 vols in one, 1697).

CARPENTER, W. B. (1839) *Principles of General and Comparative Physiology.* Churchill, London. (3rd edition, 1851).

CARPENTER, W. B. (1860) Darwin on the Origin of Species. *National Review* 10: 188–214.

CARPENTER, W. B. (1888) *Nature and Man: Essays Scientific and Philosophical.* Kegan Paul, London.

CARROLL, R. L. (1997) *Patterns and Processes of Vertebrate Evolution.* Cambridge University Press.

CARROLL, S. B. (2005) *Endless Forms Most Beautiful : The New Science of Evo Devo and the Making of the Animal Kingdom.* WW Norton, New York.

CASTLE, W. E. & PHILLIPS, J. C. (1911) On germinal transplantation in vertebrates. *Carnegie Institution of Washington Publication* 144: 1–26.

CASTLE, W. E. & PHILLIPS, J. C. (1914) Piebald rats and selection. *Carnegie Institution of Washington Publication* 195: 1–54.

CHAMBERS, R. (1844) *Vestiges of the Natural History of Creation.* Churchill, London. (Reprinted with an introduction by G. R. de Beer, Leicester University Press, 1969).

CHAMBERS, R. (1994) *Vestiges of the Natural History of Creation and Other Evolutionary Writings.* Edited by James Secord. University of Chicago Press.

COHEN, C. (2002) *The Fate of the Mammoth. Fossils, Myth, and History.* University of Chicago Press.

COLBERT, E. H., MORALES, M. & MINKOFF, E. C. (2001) *Colbert's Evolution of the Vertebrates: A History of the Backboned Animals Through Time.* 5th edn. John Wiley & Sons, New York.

CONYBEARE, W. D. (1824) On the discovery of an almost perfect skeleton of the Plesiosaurus. *Transactions of the Geological Society of London* 1: 381–9.

CONYBEARE, W. D. (1841) Letter to Lyell, see Rudwick (1967).

CONYBEARE, W. D. & PHILLIPS, W. (1822) *Outlines of the Geology of England and Wales.* Vol. 1. London.

COPE, E. D. (1887) *The Origin of the Fittest. Essays on Evolution.* Appleton, New York.

CRONIN, H. (1991) *The Ant and the Peacock. Altruism and Sexual Selection from Darwin to Today.* Cambridge University Press.

CUVIER, G. (1805) *Lecons d'Anatomie Comparee.* 5 vols. Paris.

CUVIER, G. (1806) Sur les elephans vivans et fossiles. *Annales du Museum national d'Histoire naturelle* 5: 1–58, 93–155, 249–69.

CUVIER, G. (1812) *Recherches sur les Ossemens Fossiles de Quadrupedes.* 4 vols. Chez Deterville, Paris.

CUVIER, G. (1813) *Essay on the Theory of the Earth.* Translated by R. Kerr, with notes by R. Jameson. William Blackwood, Edinburgh.

CUVIER, G. (1817) *Le Regne Animal Distribue d'apres son Organisation.* 4 vols. Paris.

CUVIER, G. (1830) Considerations sur les mollusques, et en particular sur les cephalopodes. *Annales de Sciences Naturelle* 19: 241–59.

CUVIER, G. (1997) For new translations of texts on fossils and geology, see Rudwick (1997).

DARWIN, C. (1845) *Journal of Researches into the Natural History and Geology of the Countries Visited During the Voyage H.M.S. 'Beagle' Round the World.* 2nd edition. John Murray, London.

DARWIN, C. (1851–4) *A Monograph of the Sub-Class Cirripedia, with Figures of All the Species.* 2 vols. Ray Society, London.

DARWIN, C. (1859) *On the Origin of Species by means of Natural Selection or the Preservation of Favoured Races in the Struggle for Life.* John Murray, London. (Facsimile reprint with an introduction by E. Mayr, Harvard University Press, 1964).

DARWIN, C. (1862) *On the Various Contrivances by which British and Foreign Orchids are Fertilised by Insects, and on the Good Effects of Intercrossing.* John Murray, London.

DARWIN, C. (1868) *The Variation of Animals and Plants under Domestication.* 2 vols. John Murray, London.

DARWIN, C. (1871) *The Descent of Man and Selection in Relation to Sex.* John Murray, London.

DARWIN, C. (1958) *The Autobiography of Charles Darwin.* Edited by N. Barlow. Collins, London.

DARWIN, C. (1975) *Charles Darwin's Natural Selection: Being the Second Part of His Big Species Book Written from 1856 to 1858.* Edited by R. C. Stauffer. Cambridge University Press.

DARWIN, C. (1987) *Charles Darwin's Notebooks (1836–1844).* Edited by P. H. Barrett, P. J. Gautrey, S. Herbert, D. Kohn & S. Smith. British Museum (Natural History), London.

DARWIN, C. & WALLACE, A. R. (1858a) On the tendency of species to form varieties; and on the perpetuation of varieties and species by natural means of selection. *Journal of the Linnean Society of London (Zoology)* 3: 45–62.

DARWIN, C. & WALLACE, A. R. (1958b) *Charles Darwin and Alfred Russel Wallace: Evolution by Natural Selection.* Edited by G. R. de Beer. Cambridge University Press. (Contains reprints of Darwin's 1842 Sketch and 1844 Essay, and the Darwin & Wallace 1858 Linnean Society paper).

DARWIN, E. (1789) *The Botanic Garden, Part II, containing The Loves of Plants, a Poem with Philosophical Notes.* Lichfield, London.

DARWIN, E. (1794–6) *Zoonomia, or the Laws of Organic Life.* 2 vols. Johnson, London.

DARWIN, E. (1803) *The Temple of Nature.* Johnson, London.

DARWIN, F. (1887) *The Life and Letters of Charles Darwin, including an Autobiographical Chapter.* 3 vols. John Murray, London.

DE LA BECHE, H. T. (1830) *Sections and Views Illustrative of Geological Phaenomena.* London.

D'ORBIGNY, M. C. (1849) *Dictionnaire universel d'Histoire naturelle.* 13 vols. Paris.

DE VRIES, H. (1889) *Intracellulare Pangenesis.* Jena. (English translation by C. S. Gager, University of Chicago Press, 1910).

DE VRIES, H. (1901–3) *Die Muationstheorie. Versuche und Beobachtungen uber die Enstehung von Arten im Pflanzenreich.* 2 vols. Leipzig. (English translation by J. B. Farmer and A. D. Darbishire. London, 1910–11).

DE VRIES, H. (1905) *Species and Varieties: their Origin by Mutation.* Edited by D. T. MacDougal, University of Chicago Press.

DICE, L. R. (1947) Effectiveness of selection by owls of deer mice (*Peromyscus maniculatus*) which contrast in color with their background. *Contributions of the Laboratory of Vertebrate Biology, University of Michigan* 34: 1–20.

DOBZHANSKY, T. (1937) *Genetics and the Origin of Species.* Columbia University Press, New York (3rd edition, 1951).

ELDREDGE, N. (1986) *Time Frames. The Rethinking of Darwinian Evolution and the Theory of Punctuated Equilibria.* Heinemann, London.

ENDLER, J. A. (1986) *Natural Selection in the Wild.* Princeton University Press.

FASTOVSKY, D. E. & WEISHAMPEL, D. B. (2005) *The Evolution and Extinction of the Dinosaurs.* 2nd edition. Cambridge University Press.

FAUJAS ST-FOND, B. (1799) *Histoire Naturelle de la Montagne de Saint-Pierre de Maestricht.* Paris.

FISHER, R. A. (1927) On some objections to mimicry theory; statistical and genetic. *Transactions of the Royal Entomological Society of London* 75: 269–78.

FISHER, R. A. (1930) *The Genetical Theory of Natural Selection.* Oxford University Press.

FLEAGLE, J. G. (1999) *Primate Adaptation and Evolution.* 2nd edition. Academic Press, San Diego.

FLOWER, W. H., ed (1866) *Recent Memoirs on the Cetacea by Professors Eschridt, Reinhardt and Lilljeborg.* Ray Society, London.

FLOWER, W. H. & LYDEKKER, R. (1891) *An Introduction to the Study of Mammals Living and Extinct.* A & C Black, London.

FORTEY, R. A. (2002) *Fossils: The Key to the Past.* 3rd edition. Natural History Museum Publications, London.

FUTUYMA, D. J. (2005) *Evolution.* Sinauer Associates, Sunderland, Massachusetts.

GALTON, F. (1872) On blood relationships. *Proceedings of the Royal Society of London* 20: 394–402.

GALTON, F. (1876) A theory of heredity. *Journal of the Royal Anthropological Institute of Great Britain and Ireland* 5: 329–48.

GAUDRY, A. (1862–7) *Animaux Fossile et Geologie de l'Atique d'apres les Recherches Faites en 1855–56 et 1860 sous les Auspices de l'Academie des Sciences.* 2 vols. Savy, Paris.

GAUDRY, A. (1888) *Les Ancetres de nos Animaux dans le Temps Geologiques.* Paris.

GEOFFROY-SAINT-HILAIRE, E. (1818) *Philosophie Anatomique.* Mequignon-Marvis, Paris.

GLOGER, C. L. (1833) *Das Abandern der Vogel durch Einfluss des Klimas.* August Schulz, Breslau.

GOLDSMITH, O. (1774) *A History of the Earth and Animated Nature.* 8 vols. J. Nourse, London.

GOODRICH, E. S. (1918) On the development of the segments of the head in *Scyllium. Quarterly Journal of Microscopical Science* 63: 1–30.

GOODRICH, E. S. (1924) *Living Organisms: An Account of their Origin and Evolution.* Clarendon Press, Oxford.

GOODRICH, E. S. (1930) *Studies on the Structure and Development of Vertebrates.* Macmillan, London.

GOULD, S. J. (1987) *Time's Arrow Time's Cycle. Myth and Metaphor in the Discovery of Geological Time.* Harvard University Press, Cambridge, Massachusetts.

GOULD, S. J. (1989) *Wonderful Life. The Burgess Shale and the Nature of History.* Hutchinson Radius, London.

GOULD, S. J. (2002) *The Structure of Evolutionary Theory.* Harvard University Press, Cambridge, Massachusetts.

GRANT, P. R. (1986) *Ecology and Evolution of Darwin's Finches.* Princeton University Press (reprinted with new Preface and Afterword, 1999).

GREGORY, W. K. (1929) *Our Face from Fish to Man.* Putnam's Sons, New York. (reprinted in paperback, Capricorn Books, 1965).

GULICK, J. T. (1905) Evolution, racial and habitudinal. *Carnegie Institution of Washington Publication* 25: 1–269.

HAECKEL, E. (1866) *Generelle Morphologie der Organismen.* 2 vols. Georg Reimer, Berlin.

HAECKEL, E. (1868) *Naturliche Schopfungsgeschichte.* Georg Reimer, Berlin. (English translation: *The History of Creation. Or The Development of the Earth and its Inhabitants by the Action of Natural Causes.* 2 vols. Kegan Paul, London, 1876).

HAECKEL, E. (1879) *The Evolution of Man. A Popular Exposition of the Principal Points of Human Ontogeny and Phylogeny.* 2 vols. London.

HALDANE, J. B. S. (1932) *The Causes of Evolution.* Longmans Green, London.

HALLAM, A. (2004) *Catastrophes and Lesser Calamities: The Causes of Mass Extinctions.* Oxford University Press.

HALLAM, A. & WIGNALL, P. B. (1997) *Mass Extinctions and their Aftermath.* Oxford University Press.

HERSCHEL, J. F. W. (1831) *Preliminary Discourse on the Study of Natural Philosophy.* Longman, Rees, Orme, Browne & Green, London.

HOOKE, R. (1665) *Micrographia: or some Physiological Descriptions of Minute Bodies made by Magnifying Glasses, with Observations and Inquiries Thereupon.* London.

HOOKE, R. (1705) *The Posthumous Works of Robert Hooke M. D. etc. Containing his Cutlerian Lectures, and other Discourses read at the Meetings of the Illustrious Royal Society.* Edited by R. Waller. Smith & Walford, London.

HOOKER, J. D. (1860) *The Botany of the Antarctic Voyage of H. M. S. Discovery Ships Erebus and Terror. Part III: Flora Tasmaniae.* 2 vols. Lovell & Reeve, London.

HULL, D. L. (1973) *Darwin and his Critics: The Reception of Darwin's Theory of Evolution by the Scientific Community.* Harvard University Press, Cambridge, Massachusetts.

HUTTON, J. (1788) Theory of the earth; or an investigation of the laws observable in the composition, dissolution, and restoration of land upon the globe. *Transactions of the Royal Society of Edinburgh* 1: 209–304.

HUTTON, J. (1795) *Theory of the Earth, with Proofs and Illustrations.* 2 vols. Edinburgh.

HUXLEY, T. H. (1863) *Evidence as to Man's Place in Nature.* Williams & Norgate, London.

HUXLEY, T. H. (1868) On the animals which are most nearly intermediate between birds and reptiles. *Geological Magazine* 5: 357–65. (Reprinted in *The Scientific Memoirs of Thomas Henry Huxley,* edited by M. Foster and E. R. Lankester, vol. 3: 303–13. Macmillan, London, 1898–1902).

JEPSEN, G. L., MAYR, E. & SIMPSON, G. G., eds (1949) *Genetics, Paleontology and Evolution.* Princeton University Press.

JOHANNSEN, W. (1911) The genotype conception of heredity. *American Naturalist* 45: 129–59.

JOHANSON, D. C. & EDEY, M. A. (1981) *Lucy: The Beginnings of Humankind.* Granada, London.

JONES, S., MARTIN, R. & PILBEAM, D., eds (1992) *The Cambridge Encyclopedia of Human Evolution.* Cambridge University Press.

JORDAN, D. S. & KELLOGG, V. L. (1907) *Evolution and Animal Life.* Appleton, New York. (Reprinted, 1922).

JULIN, C. (1880) Recherches sur l'ossification du maxillaire inferieur et sur la constitution du systeme dentaire chez le foetus de la *Balaenoptera rostrata. Archives de Biologie* 1: 75–136.

KEMP, T. S. (1999) *Fossils and Evolution.* Oxford University Press.

KEMP, T. S. (2004) *The Origin and Evolution of Mammals.* Oxford University Press.

KING, W. (1864) The reputed fossil man of the Neanderthal. *Quarterly Journal of Science* 1: 88—97.

KIRCHER, A. (1675) *Arca Noe in tres libros digesta, sive de rebus ante diluvium, de diluvio et de rebus post diluvium a Noemo gestis.* Waesberg, Amsterdam.

KREBS, J. R. & DAVIES, N. B. (1987) *An Introduction to Behavioural Ecology.* 2nd edition, Blackwell, Oxford.

KREBS, J. R. & DAVIES, N. B., eds (1991) *Behavioural Ecology, an Evolutionary Approach.* 3rd edition, Blackwell, Oxford.

LACK, D. (1947) *Darwin's Finches.* Cambridge University Press. (Reprinted with an introduction and notes by L. Ratcliffe & P. T. Boag, 1983).

LACK, D. (1957) *Evolutionary Theory and Christian Belief.* Methuen, London.

LAMARCK, J. B. (1778) *La Flore Francaise.* Paris.

LAMARCK, J. B. (1801) *Systeme des Animaux sans Vertebres.* Chez Deterville, Paris.

LAMARCK, J. B. (1809) *Philosophie Zoologique.* Chez Dentu, Paris. (English translation by H. Elliot, Macmillan, London, 1914; reprinted, University of Chicago Press, 1984).

LAMARCK, J. B. (1815—22) *Histoire Naturelle des Animaux sans Vertebres.* 7 vols. Paris.

LARSEN, J. L. (1994) *Interpreting Nature. The Science of Living Form from Linnaeus to Kant.* John Hopkins University Press, Baltimore.

LEAKEY, L. S. B., TOBIAS, P. V. & NAPIER, J. R. (1964) A new species of the genus Homo from Olduvai Gorge. *Nature* 202: 7—9.

LE GROS CLARK, W. E. (1950) Hominid characters of the australopithecine dentition. *Journal of the Royal Anthropological Institute of Great Britain and Ireland* 80: 37—54.

LE GROS CLARK, W. E. (1959) The Crucial Evidence for Human Evolution (Penrose Memorial Lecture). *Proceedings of the American Philosophical Society* 103: 159—72.

LE GROS CLARK, W. E. (1964) *The Fossil Evidence for Human Evolution.* 2nd edition. University of Chicago Press.

LEVINGTON, J. (2001) *Genetics, Paleontology and Evolution.* 2nd edition. Cambridge University Press.

LEWIN, R. (2004) *Human Evolution: An Illustrated Introduction.* 5th edition. Blackwell Science. Malden, Massachusetts.

LEWIS, C. L. E. & KNELL, S. J., eds (2001) *The Age of the Earth. From 4004 BC to AD 2002.* Geological Society, London, Special Publications 190.

LINNAEUS, C. (1735) *Systema Naturae.* Laurentii Salvii, Holmiae Stockholm (10th edition, 1758).

LINNAEUS, C. (1749—69) *Amoenitates academicae.* 7 vols. Laurentii Salvii, Holmiae Stockholm. (Includes *Essay on Oeconomy of Nature*, by I. Biberg, translated in 'Miscellaneous Tracts relating to Natural History … with notes by Benj. Stillingfleet', No. 2, pp. 31—108. R. & J. Dodsley & Co., London, 1759. Also *On the increase of the habitable earth*, in 'Select Dissertations from Amoenitates Academicae, A supplement to Mr Stillingfleet's Tracts …', Translated by F. J. Brand', Vol. 1, pp. 71—127. Robinson & Robinson, London, 1781).

LULL, R. S. (1929) *Organic Evolution.* Revised edition. Macmillan, New York.

LYELL, C. (1830—33) *Principles of Geology: Being an Attempt to Explain the Former Changes of the Earth's Surface by Reference to Causes Now in Operation.* 3 vols, London.

LYELL, C. (1851) *A Manual of Elementary Geology.* John Murray, London.

LYELL, C. (1863) *The Geological Evidences of the Antiquity of Man.* John Murray, London.

LYELL, C. (1970) *Sir Charles Lyell's Journals on the Species Question.* Edited by L. J. Wilson. Yale University Press, New Haven.

MALTHUS, T. R. (1826) *An Essay on the Principle of Population*. 6th edition. Johnson, London. (Darwin read this edition; 1st edition, 1798).

Marsh, O. C. (1877) Introduction and succession of vertebrate life in America. *Nature* 16: 448–50, 470–2, 489–91.

MAYR, E. (1942) *Systematics and the Origin of Species from the Viewpoint of a Zoologist*. Columbia University Press, New York.

MAYR, E. (1950) Taxonomic categories in fossil hominids. *Cold Spring Harbor Symposia in Quantitative Biology* 15: 109–18.

MAYR, E. (1963) *Animal Species and Evolution*. Harvard University Press, Cambridge, Massachusetts.

MAYR, E. (1982) *The Growth of Biological Thought. Diversity, Evolution and Inheritance*. Harvard University Press, Cambridge, Massachusetts.

MAYR, E. & PROVINE, W. B. (1980) *The Evolutionary Synthesis. Perspectives on the Unification of Biology*. Harvard University Press, Cambridge, Massachusetts. (Reprinted with new Preface by Mayr, 1998).

MENDEL, G. J. (1965) *Experiments on Plant Hybridization*. Translated with a Foreword by P. C. Mangelsdorf. Harvard University Press, Cambridge, Massachusetts.

MEYER, H. VON (1862) On the *Archaeopteryx lithographica*, from the Lithographic Slate of Solenhofen. *Annals and Magazine of Natural History* 9: 366–70.

MIVART, ST G. J. (1871) *On the Genesis of Species*. Macmillan, London.

MORGAN, T. H. (1903) *Evolution and Adaptation*. Macmillan, New York.

MORGAN, T. H. (1919) *The Physical Basis of Heredity*. Lippincott, Philadelphia.

MURCHISON, R. I. (1839) *The Silurian System*. London.

Newton, I. (1960) *The Correspondence of Isaac Newton 1676–1687*. Edited by H. W. Turnbull. Cambridge University Press.

NICHOLSON, H. A. (1879) *A Manual of Palaeontology for the Use of Students*. 2nd edition, 2 vols. William Blackwood, Edinburgh.

OSBORN, H. F. (1917) *The Origin and Evolution of Life*. Charles Scribner's Sons, New York.

OSGOOD, W. H. (1909) Revision of the mice of the American genus *Peromyscus*. *United States Department of Agriculture, North American Fauna* 28: 1–285.

OWEN, R. (1848) *On the Archetype and Homologies of the Vertebrate Skeleton*. J. van Voorst, London.

OWEN, R. (1849) *On the Nature of Limbs*. J. van Voorst, London.

OWEN, R. (1863) On the *Archaeopteryx* of von Meyer, with a description of the fossil remains of a long-tailed species, from the lithographic stone of Solenhofen. *Philosophical Transactions of the Royal Society of London* 153: 33–47.

OWEN, R. (1865) *Memoir on the Gorilla (Troglodytes gorilla, Savage)*. Taylor & Francis, London.

PHILLIPS, J. (1841) *Figures and Descriptions of the Palaeozoic Fossils of Cornwall*, Devon and West Somerset. London.

PHILLIPS, J. (1860) *Life on Earth: its Origin and Succession*. Macmillan.

PICTET, F.-J. (1844–46) *Traite de Paleontologie, ou Histoire naturelle des animaux fossiles consideres dans leurs rapport zoologiques et geologiques*. Paris.

PICTET, F.-J. (1860) Sur l'Origine de l'Espece, par Charles Darwin. *Archives des Sciences physiques et naturelles de la Bibliotheque Universelle* 3: 231–55. (English translation and notes by Hull, 1973, listed above).

PLOT, R. (1677) *The Natural History of Oxfordshire, Being an Essay toward the Natural History of England*. Oxford. (2nd edition, Brome, London 1705).

PORTER, R. (2000) *Enlightenment: Britain and the Creation of the Modern World*. Allen Lane, London. (Paperback, Penguin Books, 2001).

POULTON, E. B. (1890) *The Colours of Animals, their Meaning and Use Considered Especially in the Case of Insects*. 2nd edition. Kegan Paul, London.

PUNNETT, R. C. (1915) *Mimicry in Butterflies*. Cambridge University Press.

RAUP, D. M. (1991) *Extinction: Bad Genes or Bad Luck?* W. W. Norton, New York.

RAUP, D. M. & SEPKOSKI, J. J. (1982) Mass extinctions in the marine fossil record. *Science* 215: 1501–3.

RAY, J. (1660) *Catalogus Plantarum circa Cantabrium nascentium*. J. Field, Cambridge.

RAY, J. (1678) *The Ornithology of Francis Willughby*. John Martyn, London.

RAY, J. (1686–1704) *Historia Plantarum*. 3 vols. Henry Faithorne, London.

RAY, J. (1690) *Synopsis Methodica Stirpum Britannicum*. London. (3rd edition with illustrations, 1724; reprinted with an introduction by W. T. Stearn, Ray Society, 1973).

RAY, J. (1691) *The Wisdom of God Manifested in the Works of Creation*. Samuel Smith, London.

RAY, J. (1692) *Miscellaneous Discourses Concerning the Dissolution and Changes of the World*. Samuel Smith, London. (2nd edition, 1693, was entitled *Three Physico-Theological Discourses* and included illustrations).

REZNICK, D. N., SHAW, F. H., RODD, F. H. & SHAW, R. G. (1997) Evaluation of the rate of evolution in natural populations of guppies (*Poecilia reticulata*). *Science* 275: 1934–7.

RIDLEY, M. (2003) *Evolution*. 3rd edition. Blackwell Science, Oxford.

ROMANES, G. J. (1892) *Darwin and after Darwin. An Exposition of the Darwinian Theory and a Discussion of post-Darwinian Questions. I. The Darwinian Theory*. Longmans Green, London.

ROMER, A. S. (1970) The Chanares (Argentina) Triassic reptile fauna. VI. A chiniquodontid cynodont with an incipient squamosal–dentary jaw articulation. *Breviora* 344: 1–18.

ROTHSCHILD, LORD & DOLLMAN, G. (1936) The genus Dendrolagus. *Transactions of the Zoological Society of London* 21: 477–548.

RUDWICK, M. J. S. (1967) A critique of uniformitarian geology: a letter from W. D. Conybeare to Charles Lyell, 1841. *Proceedings of the American Philosophical Society* 111: 272–87.

RUDWICK, M. J. S. (1976) *The Meaning of Fossils. Episodes in the History of Palaeontology*. 2nd edition. University of Chicago Press.

RUDWICK, M. J. S. (1992) *Scenes from Deep Time: Early Pictorial Representations of the Prehistoric World*. University of Chicago Press.

RUDWICK, M. J. S. (1997) *Georges Cuvier, Fossil Bones and Geological Catastrophes: New Translations and Interpretations of the Primary Texts*. University of Chicago Press.

RUSE, M. (1979) *The Darwinian Revolution. Science Red in Tooth and Claw*. University of Chicago Press. (Reprinted with new Afterword, 1999).

SARICH, V. & WILSON, A. C. (1967) Immunological time-scale for human evolution. *Science* 158: 1200–3.

SCROPE, G. P. (1827) *Memoir on the Geology of Central France, including the Volcanic Formations of Auvergne, the Velay and the Vivarais*. London.

SEBRIGHT, J. (1809) *The Art of Improving the Breeds of Domestic Animals, in a Letter Addressed to the Right Hon. Sir Joseph Banks*. K. B., London.

SECORD, J. A. (2000) *Victorian Sensation: The Extraordinary Publication, Reception, and Secret Authorship of* Vestiges of the Natural History of Creation. University of Chicago Press.

SEDGWICK, A. (1860) Objections to Mr. Darwin's theory of the origin of species. *Spectator*, 24 March.

SEPKOSKI, J. J. (1993) Ten years in the library: new data confirm paleontological patterns. *Paleobiology* 19: 43–51.

SIBLEY, C. G. & AHLQUIST, J. E. (1984) The phylogeny of the hominid primates, as indicated by DNA–DNA hybridisation. *Journal of Molecular Biology* 20: 2–15.

SIBLEY, C. G. & AHLQUIST, J. E. (1987) DNA hybridisation evidence of hominoid phylogeny: results from an expanded data set. *Journal of Molecular Biology* 26: 99–121.

SIMPSON, G. G. (1944) *Tempo and Mode in Evolution.* Columbia University Press, New York.

SIMPSON, G. G. (1949) *The Meaning of Evolution.* Yale University Press, New Haven. (Revised edition, 1967).

SIMPSON, G. G. (1951) *Horses.* Oxford University Press, New York.

SIMPSON, G. G. (1953) *The Major Features of Evolution.* Colombia University Press, New York.

SMITH, W. (1816) *Strata Identified by Organised Fossils.* London.

STANLEY, S. M. (1981) *The New Evolutionary Timetable. Fossils, Genes and the Origin of Species.* Basic Books, New York.

STENO, N. (1669) *De Solido intra Solidum naturaliter Contento Dissertationis Prodromus.* Printing Shop, Florence.

STENO, N. (1671) *The Prodromus to a Dissertation Concerning Solids Naturally Contained within Solids.* Translated by H. Oldenburg. Moses Pitt, London. (For a modern English translation, see *Steno, geological papers*, translated A. J. Pollock, edited by G. Scherz, Odense University Press, 1969).

SUMNER, F. B. (1930) Genetic and distributional studies of three sub-species of Peromyscus. *Journal of Genetics* 23: 275–376.

SUMNER, F. B. (1932) Genetic, distributional and evolutionary studies of the subspecies of deermice (*Peromyscus*). *Bibliographia Genetica* 9: 1–106.

SUMNER, F. B. (1945) *The Life History of an American Naturalist.* Jaques Cattell Press, Lancaster, Pennsylvania.

STRINGER, C. & ANDREWS, P. (2005) *The Complete World of Human Evolution.* Thames & Hudson, The Natural History Museum, London.

TATTERSALL, I. (1995) *The Fossil Trail: How We Know What We Think We Know about Human Evolution.* Oxford University Press, New York.

THOMAS, K. (1983) *Man and the Natural World. Changing Attitudes in England 1500–1800.* Allen Lane, London. (Reprinted in paperback, Penguin Books, 1984).

TOPSELL, E. (1607) *The Historie of Four-footed Beasts and Serpents.* London (Reprinted 1658).

TOULMIN, S. & GOODFIELD, J. (1965) *The Discovery of Time.* Hutchinson, London. (Reprinted in paperback, Penguin Books, 1967).

TYSON, E. (1699) *Orang-Outang, sive Homo Sylvestris.* London.

USSHER, J. (1650) *Annales Veteris Testamenti, a prima Mundi Origine Deducti.* London.

WALLACE, A. R. (1855) On the law which has regulated the introduction of new species. *Annals and Magazine of Natural History* 16: 184–96. (Reprinted in Wallace, 1891).

WALLACE, A. R. (1869) *The Malay Archipelago: the Land of the Orang-utan, and the Bird of Paradise.* Macmillan, London. (Reprinted in paperback, with an introduction by J. Bastin, Oxford University Press, 1989).

WALLACE, A. R. (1876) *The Geographical Distribution of Animals.* 2 vols. Macmillan, London.

WALLACE, A. R. (1889) *Darwinism.* Macmillan, London.

WALLACE, A. R. (1891) *Natural Selection and Tropical Nature. Essays on Descriptive and Theoretical Biology.* Macmillan, London.

WALLACE, A. R. (1905) *My Life: A Record of Events and Opinions*. Chapman and Hall, London.

WATSON, J. D. (1965) *Molecular Biology of the Gene*. Benjamin, New York.

WEISMANN, A. (1892) *Das Keimplasma: eine Theorie der Vererbung*. Gustav Fischer, Jena. (English translation by W. N. Parker and H. Ronnfeldt: *The Germ-Plasm: a Theory of Heredity*. Walter Scott, London, 1893).

WEISMANN, A. (1904) *The Evolution Theory*. 2 vols. Translated with the author's co-operation by J. A. Thomson and M. R. Thomson. Edward Arnold, London.

WERNER, A. G. (1787) *Kurze Klassifikation und Beschreibung der verschiedenen Gebirgsarten*. Freiberg. (English translation by A. M. Ospovat: *Short Classification and Description of the Various Rocks*, Hafner Publishing, New York, 1971).

WOODWARD, J. (1695) *An Essay toward a Theory of the Earth and Terrestrial Bodies, especially Minerals; as also of the Seas, Rivers and Springs*. Wilkin, London.

WRIGHT, S. (1931) Evolution in Mendelian populations. *Genetics* 16: 97—159.

ILLUSTRATION SOURCES

COLOUR

Block 1

Plate 1A. *Macropus giganteus* by George Raper. Copyright Natural History Museum, London.

Plate 1B. *Dendrolagus lumholtzi*. From Rothschild, Lord & Dollman, G. (1936) *Transactions of the Zoological Society*, Vol. 21. Copyright Natural History Museum, London.

Plate 2A & B. Orang-utan and chimpanzee. From Cuvier, G. (1817) *Le Regne Animal*. Copyright Natural History Museum, London.

Plate 3A. Hyaena in cemetry. From MS Bodley 764, fol. 15r, with permission of The Bodleian Library, University of Oxford.

Plate 3B. *Hyaena vulgaris*. From D'Orbigny, C. (1849) *Dictionnaire Universel d'Histoire Naturelle*. Copyright Natural History Museum, London.

Plate 4. Page of plants. From Culpeper, N. (1649) *The Complete Herbal*. Copyright Natural History Museum, London.

Block 2

Plate 5A. Lady's slipper orchid by Franz Bauer. Copyright Natural History Museum, London.

Plate 5B. Title page. From Linnaeus, C. (1758) *Systema Naturae*, 10th edition. Copyright Natural History Museum, London.

Plate 6A. Village and volcano. From Scrope, G. P. (1827) *Memoir on the Geology of Central France*. Copyright Natural History Museum, London.

Plate 6B. Bove Valley. From Lyell, C. (1832) *Principles of Geology*, Volume II. Copyright Natural History Museum, London.

Plate 7. Secondary Series. From Buckland, W. (1836) *Geology and Mineralogy considered with reference to Natural Theology*. Copyright Natural History Museum, London.

Plate 8A. Embryonic whale teeth. From Julin, C. (1880) *Archives de Biologie*, Vol.1. Copyright Natural History Museum, London.

Plate 8B. Vertebrate and cephalopod. From Cuvier, G. (1830) *Annales de Sciences Naturelle*, Vol. 19. Copyright Natural History Museum, London.

Block 3

Plate 9. Photographs of *Geospiza fortis*, *G. magnirostris* and *G. fuliginosa* specimens. Copyright Natural History Museum. London.

Plate 10. *Eucalyptus urnigera*. From Hooker, J. D. (1860) *The Botany of the Antarctic Voyage etc. Part III, Flora Tasmaniae*. Copyright Natural History Museum, London.

Plate 11. Butterfly mimicry. From Bates, H. W. (1862) On the lepidoptera of the Amazon valley. *Transactions of the Linnean Society*, Vol. 23. Copyright Natural History Museum, London.

Plate 12A. Caterpillar warning colours. From Buckler, W. (1886—1901) *The Larvae of the British Butterflies and Moths*. Copyright Natural History Museum, London.

Plate 12B. Moth colour patterns. From Bateson, W. (1913) *Mendel's Principles of Heredity*. Copyright Natural History Museum, London.

Block 4

Plate 13. *Achatinella*. From Gulick, J. T. (1905) Evolution, Racial and Habitudinal. *Carnegie Institution of Washington Publication* 25. Copyright Natural History Museum, London.

Plate 14. Map of *Peromyscus maniculatus* distribution. From Osgood, W. H. (1909) Revision of the mice of the American genus *Peromyscus*. *United States Department of Agriculture, North American Fauna*, No. 28. Copyright Natural History Museum, London.

Plate 15. Mouse skins. From Sumner, F. B. (1930) Genetic and distributional studies of three sub-species of *Peromyscus*. *Journal of Genetics*, Vol. 23. Copyright Natural History Museum, London.

Plate 16A. Photograph of *Australopithecus afarensis* skeleton (Lucy). Copyright Natural History Museum, London.

Plate 16B. Photograph of *Homo erectus* (Turkana Boy) skull. Copyright Natural History Museum, London.

BLACK AND WHITE

Chapter 1

p. 5. Kangaroo. From Flower, W. H. & Lydekker, R. (1891) *An Introduction to the Study of Mammals*. Copyright Natural History Museum, London.

p. 8. Hind feet of marsupials. From Lull, R. S. (1929) *Organic Evolution*. Copyright Natural History Museum, London.

p. 10. Introduction to Darwin/Wallace papers. From *Linnean Society Proceedings* (1858). Copyright Natural History Museum, London.

p. 12. Discovery of mosasaur jaw. From Faujas de Saint Fond, F. (1799) *Histoire Naturelle de al Montagne de St Pierre de Maestricht*. Copyright Natural History Museum, London.

p. 13. Mastodon engraving. Copyright Natural History Museum, London.

p. 14. Adult gorilla and offspring. From Owen, R. (1865) *Memoir on the Gorilla*. Copyright Natural History Museum, London.

p. 17. Fish resembling a monk. From Belon, P. (1553) *De Aquatilibus*. Copyright Natural History Museum, London.

p. 18. Human and bird skeletons. From Belon, P. (1555) *Histoire de la nature des Oiseux*. Copyright Natural History Museum, London.

Chapter 2

p. 21. *Potamogeton*. From Ray, J. (1724) *Synopsis Methodica Stirpium Britannicarium*. Copyright Natural History Museum, London.

p. 23. Woodcut of bison. From Topsell, E. (1658) *Historie of Four-Footed Beasts and Serpents*. Copyright Natural History Museum, London.

p. 26. Owls. From Ray, J. (1678) *The Ornithology of Francis Willughby*. Copyright Natural History Museum, London.

p. 30. Woodpeckers. From Ray, J. (1678) *The Ornithology of Francis Willughby*. Copyright Natural History Museum, London.

p. 32. House fly. From Hooke, R. (1665) *Micrographia*. Copyright Natural History Museum, London.

p. 35. Dinosaur bone. From Plot, R. (1677) *The Natural Historie of Oxfordshire*. Copyright Natural History Museum, London.

p. 37. Ammonites. From Hooke, R. (1705) *The Posthumous Works of Robert Hooke*. Ed. R. Waller. Copyright Natural History Museum, London.

p. 40. Frontispiece. From Burnet, T. (1690) *Sacred Theory of the Earth*. Copyright Natural History Museum, London.

p. 42. Fossil fish. From Ray, J. (1693) *Three Physico-Theological Discourses*. Copyright Natural History Museum, London.

p. 44. Drawing of chimp skeleton. From Tyson, E. (1699) *Orang-Outang sive Homo sylvestris*. Copyright Natural History Museum, London.

p. 44. Photo of Tyson's chimp. Copyright Natural History Museum, London.

Chapter 3

p. 49. *Vallisneria spiralis*. From Darwin, E. (1789) *The Botanic Garden, a poem*. Copyright Natural History Museum, London.

p. 51. Four "manlike apes". From *Selecta Disserta Linnaei*, Vol. III. Copyright Natural History Museum, London.

p. 52. Noah's Ark. From Kircher, A. (1675) *Arca Noe*. Copyright Natural History Museum, London.

p. 55. Tarsier. From Buffon, G. (1785) *Natural History*, English translation by William Smellie, Vol. 7. Copyright Natural History Museum, London.

p. 57. Drawings of big cats. From Bewick, T. (1885) *Thomas Bewick's Works, Vol. III History of Quadrupeds*. Copyright Natural History Museum, London.

p. 60. Three elephant-like femurs. From Daubenton, L. (1762) *Memoires de l'Academie des Sciences*, Vol. 228. Copyright Natural History Museum, London.

p. 63. Geological section. Drawing by Sally Alexander. Copyright Sally Alexander/Natural History Museum, London.

p. 66. Trap dyke & cliff erosion. From De La Beche, H. T. (1830) *Sections and Views Illustrative of Geological Phaenomena*. Copyright Natural History Museum, London.

p. 67. Drawing and diagram of Siccar Point. From Lyell, C. (1852) *Elements of Geology*. Copyright Natural History Museum, London.

p. 70. *Megatherium*. From Cuvier, G. (1817) *Le Regne Animal*, Vol. 2. Copyright Natural History Museum, London.

p. 71. Three elephant molars. From Blumenbach, J. F. (1799) *Handbuch fur Naturgeschichte*, 6th ed. Copyright Natural History Museum, London.

p. 73. *Palaeotherium*. From Buckland, W. (1836) *Geology and Mineralogy Considered with Reference to Natural Theology*. Copyright Natural History Museum, London.

p. 74. *Darwinia lejostyla*. Drawing by Peter Neish. Courtesy of David Young.

Chapter 4

p. 81. Shoulder bones. From Geoffroy-Saint-Hilaire, E. (1818) *Philosophie Anatomique*. Copyright Natural History Museum, London.

p. 82. Dissection of octopus eye. From Cuvier, G. (1817) *Memoires pour servir a l'histoire et l'anatomie des mollusques*. Copyright Natural History Museum, London.

p. 84. Cretaceous fossils. From Smith, W. (1816) *Strata Identified by Organised Fossils*. Copyright Natural History Museum, London.

p. 86. Buckland lecturing. Courtesy of the Oxford University Museum of Natural History.

p. 88. *Plesiosaurus dolichodeirus*. From Buckland, W. (1836) *Geology and Mineralogy Considered with Reference to Natural Theology*. Copyright Natural History Museum, London.

p. 89. Photograph of *Plesiosaurus hawkinsii*. Copyright Natural History Museum, London.

p. 91. Valley in Auvergne. From Lyell, C. (1852) *Elements of Geology*. Copyright Natural History Museum, London.

p. 94. Miocene fossil shells. From Lyell, C. (1833) *Principles of Geology*, Vol. III. Copyright Natural History Museum, London.

p. 97. Glacier with moraines. From Wallace, A. R. (1892) *Island Life*, 2nd ed. Copyright Natural History Museum, London.

p. 99. *Cephalaspis*. From Murchison, R. I. (1839) *The Silurian System*. Copyright Natural History Museum, London.

p. 100. Fish diversity through time. From Agassiz, L. (1833–44) *Recherches sur les Poissons fossiles*. Copyright Natural History Museum, London.

p. 101. Table of geological strata. From Lyell, C. (1863) *The Antiquity of Man*. Copyright Natural History Museum, London.

Chapter 5

p. 103. Title page. From Chambers, R. (1969) *Vestiges of the Natural History of Creation*, Leicester University Press reprint, with permission from the University of Leicester.

p. 109. Galápagos Islands. From Keynes, R. D. (1988) *Charles Darwin's Beagle Diary*, Cambridge University Press, with permission of the publisher.

p. 110. Galápagos Finches. From Darwin, C. (1884) *Journal of Researches ... during Voyage of H. M. S. Beagle*. 2nd edition. Copyright Natural History Museum, London.

p. 111. Tree diagram of evolution. From Darwin, C. (1837) *First Notebook on Transmutation of Species*. Copyright University of Cambridge Library, used with permission.

p. 115. Skeletons of dugong and bat. From Owen, R. (1849) *On the Nature of Limbs*. Copyright Natural History Museum, London.

p. 117. Tree diagram of evolution. From Bronn, H. (1861) Essai. *Supplement aux Comptes Rendus des Seances de l'Academie des Sciences*, Vol. 2. Courtesy of the University of Melbourne.

p. 119. Skull of babirusa. From Wallace, A. R. (1869) *The Malay Archipelago*. Copyright Natural History Museum, London.

p. 122. *Chthamalus* species. From Darwin, C. (1854) *A Monograph of the Sub-Class Cirripedia*, Vol. 2. Copyright Natural History Museum, London.

p. 125. English Pouter. From Darwin, C. (1905) *Variation of Animals and Plants under Domestication*, Vol. 1. Courtesy of the University of Melbourne.

p. 127. *Mylodon*. From Nicholson, H. A. (1879) *A Manual of Palaeontology*, Vol. 2. Courtesy of the University of Melbourne.

p. 129. Whale skeleton. From Flower, W. H. ed. (1866) *Recent memoires on the Cetacea by Professors Eschridt, Reinhardt and Lilljeborg*. Courtesy of the University of Melbourne.

Chapter 6

p. 131. Title page. From Darwin, C. (1859) *The Origin of Species*. Copyright Natural History Museum, London.

p. 135. Animals of Borneo and New Guinea. From Wallace, A. R. (1876) *The Geographical Distribution of Animals*. Vol. I. Copyright Natural History Museum, London.

p. 138. *Archaeopteryx*. From Romanes, G. J. (1897) *Darwin & after Darwin*, 2nd edition. Copyright Natural History Museum, London.

p. 139. Foot bones. From Nicholson, H. A. (1879) *A Manual of Palaeontology*, 2nd edition, Vol. 2. Copyright Natural History Museum, London.

p. 141. *Phenacodus*. From Cope, E. D. (1887) *The Origin of the Fittest*. Copyright Natural History Museum, London.

p. 142. Evolutionary tree. From Haeckel, E. (1879) *The Evolution of Man*. Copyright Natural History Museum, London.

p. 143. Vestigial tail bones. From Romanes, G. J. (1897) *Darwin & after Darwin*, 2nd edition. Copyright Natural History Museum, London.

p. 145. Acheulean hand axe. From Lyell, C. (1863) *Antiquity of Man*, 2nd edition. Copyright Natural History Museum, London.

p. 147. Neanderthal man remains. From Jordan, D. & Kellog, V. (1907) *Evolution and Animal Life*. Copyright Natural History Museum, London.

p. 152. Leaf insect. From Mivart, St. G. (1871) *The Genesis of Species*, 2nd edition. Copyright Natural History Museum, London.

p. 154. Diagram of variation. From Wallace (1890) *Darwinism*, 2nd edition. Copyright Natural History Museum, London.

Chapter 7

p. 159. Continuity of the germ-plasm. From Goodrich, E. (1924) *Living Organisms*. Copyright Natural History Museum, London.

p. 163. Hind feet of cat. From Bateson, W. (1894) *Materials for the Study of Variation*. Copyright Natural History Museum, London.

p. 165. Mendelian ratios. From Bateson, W. (1913) *Mendel's Principles of Heredity*. Copyright Natural History Museum, London.

p. 167. *Dahlia variabilis fistulosa*. From De Vries, H. (1910) *The Mutation Theory*, translated by J. B. Farmer & A. D. Darbyshire. Copyright Natural History Museum, London.

p. 170. Pure line of beans. From Johannsen, W. (1911) *American Naturalist*, Vol. XLV. Copyright Natural History Museum, London.

p. 171. *Drosophila*. From Morgan, T. H. (1919) *The Physical Basis of Heredity*. Copyright Natural History Museum, London.

p. 175. Head segments in dogfish. From Goodrich, E. S. (1918) *Quarterly Journal of Microscopical Science*, Vol. 63. Courtesy of the University of Melbourne.

p. 177. Fish & amphibian skulls. From Gregory, W. K. (1929) *Our Face from Fish to Man*. Copyright Natural History Museum, London.

p. 178. Photograph of *Eryops* skeleton. Copyright Natural History Museum, London.

p. 180. Mammal-like reptile skulls. From Gregory, W. K. (1929) *Our Face from Fish to Man*. Copyright Natural History Museum, London.

p. 182. Adaptive radiation in ungulates. From Goodrich, E. (1924) *Living Organisms*. Copyright Natural History Museum, London.

Chapter 8

p. 184. Norton's table of figures. From Punnett, R. C. (1915) *Mimicry in Butterflies*. Copyright Natural History Museum, London.

p. 187. Coat patterns in piebald rats. From Castle, W. E. & Phillips, J. C. (1914) Piebald rats and selection. *Carnegie Institution of Washington Publication* 195. Copyright Natural History Museum, London.

p. 191. Branching evolution. From Stauffer, R. C. (1975) *Charles Darwin's Natural Selection*, Cambridge University Press, with permission of the publisher.

p. 193. Map of Oahu. From Gulick, J. T. (1905) Evolution, Racial and Habitudinal. *Carnegie Institution of Washington Publication* 25. Copyright Natural History Museum, London.

p. 196. Inheritance of *Peromyscus* coat colour. From *Genetics and the Origin of Species* by T. Dobzhansky. Copyright (1937) Columbia University Press, used with permission of the publisher.

p. 197. Geographical variation in *Peromyscus*. From Sumner, F. B. (1932) *Bibliographia Genetica*, Vol. 9. Copyright Natural History Museum, London.

p. 200. Map of scale insect distribution. From *Genetics and the Origin of Species*, 3rd edition, by T. Dobzhansky. Copyright (1951) Columbia University Press, used with permission of the publisher.

p. 203. Distribution of *Tanysiptera* in New Guinea. From Mayr, E. (1963) *Animal Species and Evolution*, Harvard University Press, with permission of the publisher (originally from Mayr, in Huxley *et al.*, eds, 1954, *Evolution as a Process*, Allen and Unwin).

p. 205. Darwin's finches. From Lack, D. (1983) *Darwin's Finches*, Cambridge University Press, with permission of the publisher.

p. 208. Horse evolution. From *Horses*, by G. G. Simpson, copyright 1951 by Oxford University Press, Inc. Used by permission of Oxford University Press, Inc.

Chapter 9

p. 211. DNA double helix. From *Molecular Biology of the Gene*, by J. D. Watson (1965). Used by permission of Addison Wesley Longman Publishers, Inc.

p. 215. Hominid phylogeny. From Jones, S. *et al.*, eds (1992) *Cambridge Encyclopedia of Human Evolution*, Cambridge University Press, with permission of the publisher.

p. 217. *Probainognathus* skull. From Romer, A. S. (1970) The Chanares (Argentina) reptile fauna VI. *Breviora*, Vol. 344. Copyright Museum of Comparative Zoology, Harvard University, used with permission.

p. 219. Evolution of hominid skull. From Le Gros Clark, W. E. (1959) *Proceedings of the American Philosophical Society*, Vol. 103, with permission of the publisher.

p. 224. Hominid phylogeny. From *Primate Adaptation and Evolution*, 2nd edition, by J. G. Fleagle, page 541. Copyright 1999, with permission from Elsevier.

p. 226. Natural selection in *Geospiza*. Reprinted with permission from Boag, P. T. and Grant, P. R., *Science*, Vol. 214, p. 83. Copyright 1981 AAAS.

p. 228. Guppies from Trinidad. Photograph by Benoni Seghers, courtesy of David Reznick.

p. 233. Graph of mass extinction. Reprinted with permission from Raup, D. M. & Sepkoski, J. J., *Science*, Vol. 215, p. 1502. Copyright 1982 AAAS.

Chapter 10

p. 237. Tree diagram of vertebrate evolution. From Haeckel, E. (1866) *Generelle Morphologie der Organismen*. Copyright Natural History Museum, London.

p. 243. Orchid flowers. From Darwin, C. (1862) *The Fertilization of Orchids*. Copyright Natural History Museum, London.

p. 244. Photograph of *Phorusrhacus* skull. Copyright Natural History Museum, London.

p. 247. Victorian cartoon of evolution. From Lack, D. (1957) *Evolutionary Theory and Christian Belief*, with permission from the University of Leicester.

INDEX